LONDON MATHEMATICAL SOCIETY LECTURE NOTE SERIES

The titles below are available from booksellers, or from Cambridge University Press at
www.cambridge.org/mathematics

287 Topics on Riemann surfaces and Fuchsian groups, E. BUJALANCE, A.F. COSTA & E. MARTÍNEZ (eds)
288 Surveys in combinatorics, 2001, J.W.P. HIRSCHFELD (ed)
289 Aspects of Sobolev-type inequalities, L. SALOFF-COSTE
290 Quantum groups and Lie theory, A. PRESSLEY (ed)
291 Tits buildings and the model theory of groups, K. TENT (ed)
292 A quantum groups primer, S. MAJID
293 Second order partial differential equations in Hilbert spaces, G. DA PRATO & J. ZABCZYK
294 Introduction to operator space theory, G. PISIER
295 Geometry and integrability, L. MASON & Y. NUTKU (eds)
296 Lectures on invariant theory, I. DOLGACHEV
297 The homotopy category of simply connected 4-manifolds, H.-J. BAUES
298 Higher operads, higher categories, T. LEINSTER (ed)
299 Kleinian groups and hyperbolic 3-manifolds, Y. KOMORI, V. MARKOVIC & C. SERIES (eds)
300 Introduction to Möbius differential geometry, U. HERTRICH-JEROMIN
301 Stable modules and the D(2)-problem, F.E.A. JOHNSON
302 Discrete and continuous nonlinear Schrödinger systems, M.J. ABLOWITZ, B. PRINARI & A.D. TRUBATCH
303 Number theory and algebraic geometry, M. REID & A. SKOROBOGATOV (eds)
304 Groups St Andrews 2001 in Oxford I, C.M. CAMPBELL, E.F. ROBERTSON & G.C. SMITH (eds)
305 Groups St Andrews 2001 in Oxford II, C.M. CAMPBELL, E.F. ROBERTSON & G.C. SMITH (eds)
306 Geometric mechanics and symmetry, J. MONTALDI & T. RATIU (eds)
307 Surveys in combinatorics 2003, C.D. WENSLEY (ed.)
308 Topology, geometry and quantum field theory, U.L. TILLMANN (ed)
309 Corings and comodules, T. BRZEZINSKI & R. WISBAUER
310 Topics in dynamics and ergodic theory, S. BEZUGLYI & S. KOLYADA (eds)
311 Groups: topological, combinatorial and arithmetic aspects, T.W. MÜLLER (ed)
312 Foundations of computational mathematics, Minneapolis 2002, F. CUCKER *et al* (eds)
313 Transcendental aspects of algebraic cycles, S. MÜLLER-STACH & C. PETERS (eds)
314 Spectral generalizations of line graphs, D. CVETKOVIĆ, P. ROWLINSON & S. SIMIĆ
315 Structured ring spectra, A. BAKER & B. RICHTER (eds)
316 Linear logic in computer science, T. EHRHARD, P. RUET, J.-Y. GIRARD & P. SCOTT (eds)
317 Advances in elliptic curve cryptography, I.F. BLAKE, G. SEROUSSI & N.P. SMART (eds)
318 Perturbation of the boundary in boundary-value problems of partial differential equations, D. HENRY
319 Double affine Hecke algebras, I. CHEREDNIK
320 L-functions and Galois representations, D. BURNS, K. BUZZARD & J. NEKOVÁŘ (eds)
321 Surveys in modern mathematics, V. PRASOLOV & Y. ILYASHENKO (eds)
322 Recent perspectives in random matrix theory and number theory, F. MEZZADRI & N.C. SNAITH (eds)
323 Poisson geometry, deformation quantisation and group representations, S. GUTT *et al* (eds)
324 Singularities and computer algebra, C. LOSSEN & G. PFISTER (eds)
325 Lectures on the Ricci flow, P. TOPPING
326 Modular representations of finite groups of Lie type, J.E. HUMPHREYS
327 Surveys in combinatorics 2005, B.S. WEBB (ed)
328 Fundamentals of hyperbolic manifolds, R. CANARY, D. EPSTEIN & A. MARDEN (eds)
329 Spaces of Kleinian groups, Y. MINSKY, M. SAKUMA & C. SERIES (eds)
330 Noncommutative localization in algebra and topology, A. RANICKI (ed)
331 Foundations of computational mathematics, Santander 2005, L.M PARDO, A. PINKUS, E. SÜLI & M.J. TODD (eds)
332 Handbook of tilting theory, L. ANGELERI HÜGEL, D. HAPPEL & H. KRAUSE (eds)
333 Synthetic differential geometry (2nd Edition), A. KOCK
334 The Navier–Stokes equations, N. RILEY & P. DRAZIN
335 Lectures on the combinatorics of free probability, A. NICA & R. SPEICHER
336 Integral closure of ideals, rings, and modules, I. SWANSON & C. HUNEKE
337 Methods in Banach space theory, J.M.F. CASTILLO & W.B. JOHNSON (eds)
338 Surveys in geometry and number theory, N. YOUNG (ed)
339 Groups St Andrews 2005 I, C.M. CAMPBELL, M.R. QUICK, E.F. ROBERTSON & G.C. SMITH (eds)
340 Groups St Andrews 2005 II, C.M. CAMPBELL, M.R. QUICK, E.F. ROBERTSON & G.C. SMITH (eds)
341 Ranks of elliptic curves and random matrix theory, J.B. CONREY, D.W. FARMER, F. MEZZADRI & N.C. SNAITH (eds)
342 Elliptic cohomology, H.R. MILLER & D.C. RAVENEL (eds)
343 Algebraic cycles and motives I, J. NAGEL & C. PETERS (eds)
344 Algebraic cycles and motives II, J. NAGEL & C. PETERS (eds)

345 Algebraic and analytic geometry, A. NEEMAN
346 Surveys in combinatorics 2007, A. HILTON & J. TALBOT (eds)
347 Surveys in contemporary mathematics, N. YOUNG & Y. CHOI (eds)
348 Transcendental dynamics and complex analysis, P.J. RIPPON & G.M. STALLARD (eds)
349 Model theory with applications to algebra and analysis I, Z. CHATZIDAKIS, D. MACPHERSON, A. PILLAY & A. WILKIE (eds)
350 Model theory with applications to algebra and analysis II, Z. CHATZIDAKIS, D. MACPHERSON, A. PILLAY & A. WILKIE (eds)
351 Finite von Neumann algebras and masas, A.M. SINCLAIR & R.R. SMITH
352 Number theory and polynomials, J. MCKEE & C. SMYTH (eds)
353 Trends in stochastic analysis, J. BLATH, P. MÖRTERS & M. SCHEUTZOW (eds)
354 Groups and analysis, K. TENT (ed)
355 Non-equilibrium statistical mechanics and turbulence, J. CARDY, G. FALKOVICH & K. GAWEDZKI
356 Elliptic curves and big Galois representations, D. DELBOURGO
357 Algebraic theory of differential equations, M.A.H. MACCALLUM & A.V. MIKHAILOV (eds)
358 Geometric and cohomological methods in group theory, M.R. BRIDSON, P.H. KROPHOLLER & I.J. LEARY (eds)
359 Moduli spaces and vector bundles, L. BRAMBILA-PAZ, S.B. BRADLOW, O. GARCÍA-PRADA & S. RAMANAN (eds)
360 Zariski geometries, B. ZILBER
361 Words: Notes on verbal width in groups, D. SEGAL
362 Differential tensor algebras and their module categories, R. BAUTISTA, L. SALMERÓN & R. ZUAZUA
363 Foundations of computational mathematics, Hong Kong 2008, F. CUCKER, A. PINKUS & M.J. TODD (eds)
364 Partial differential equations and fluid mechanics, J.C. ROBINSON & J.L. RODRIGO (eds)
365 Surveys in combinatorics 2009, S. HUCZYNSKA, J.D. MITCHELL & C.M. RONEY-DOUGAL (eds)
366 Highly oscillatory problems, B. ENGQUIST, A. FOKAS, E. HAIRER & A. ISERLES (eds)
367 Random matrices: High dimensional phenomena, G. BLOWER
368 Geometry of Riemann surfaces, F.P. GARDINER, G. GONZÁLEZ-DIEZ & C. KOUROUNIOTIS (eds)
369 Epidemics and rumours in complex networks, M. DRAIEF & L. MASSOULIÉ
370 Theory of p-adic distributions, S. ALBEVERIO, A.YU. KHRENNIKOV & V.M. SHELKOVICH
371 Conformal fractals, F. PRZYTYCKI & M. URBAŃSKI
372 Moonshine: The first quarter century and beyond, J. LEPOWSKY, J. MCKAY & M.P. TUITE (eds)
373 Smoothness, regularity, and complete intersection, J. MAJADAS & A. G. RODICIO
374 Geometric analysis of hyperbolic differential equations: An introduction, S. ALINHAC
375 Triangulated categories, T. HOLM, P. JØRGENSEN & R. ROUQUIER (eds)
376 Permutation patterns, S. LINTON, N. RUŠKUC & V. VATTER (eds)
377 An introduction to Galois cohomology and its applications, G. BERHUY
378 Probability and mathematical genetics, N. H. BINGHAM & C. M. GOLDIE (eds)
379 Finite and algorithmic model theory, J. ESPARZA, C. MICHAUX & C. STEINHORN (eds)
380 Real and complex singularities, M. MANOEL, M.C. ROMERO FUSTER & C.T.C WALL (eds)
381 Symmetries and integrability of difference equations, D. LEVI, P. OLVER, Z. THOMOVA & P. WINTERNITZ (eds)
382 Forcing with random variables and proof complexity, J. KRAJÍČEK
383 Motivic integration and its interactions with model theory and non-Archimedean geometry I, R. CLUCKERS, J. NICAISE & J. SEBAG (eds)
384 Motivic integration and its interactions with model theory and non-Archimedean geometry II, R. CLUCKERS, J. NICAISE & J. SEBAG (eds)
385 Entropy of hidden Markov processes and connections to dynamical systems, B. MARCUS, K. PETERSEN & T. WEISSMAN (eds)
386 Independence-friendly logic, A.L. MANN, G. SANDU & M. SEVENSTER
387 Groups St Andrews 2009 in Bath I, C.M. CAMPBELL *et al* (eds)
388 Groups St Andrews 2009 in Bath II, C.M. CAMPBELL *et al* (eds)
389 Random fields on the sphere, D. MARINUCCI & G. PECCATI
390 Localization in periodic potentials, D.E. PELINOVSKY
391 Fusion systems in algebra and topology, M. ASCHBACHER, R. KESSAR & B. OLIVER
392 Surveys in combinatorics 2011, R. CHAPMAN (ed)
393 Non-abelian fundamental groups and Iwasawa theory, J. COATES *et al* (eds)
394 Variational Problems in Differential Geometry, R. BIELAWSKI, K. HOUSTON & M. SPEIGHT (eds)
395 How groups grow, A. MANN
396 Arithmetic Differential Operators over the p-adic Integers, C.C. RALPH & S.R. SIMANCA
397 Hyperbolic geometry and applications in quantum Chaos and cosmology, J. BOLTE & F. STEINER (eds)
398 Mathematical models in contact mechanics, M. SOFONEA & A. MATEI
399 Circuit double cover of graphs, C.-Q. ZHANG
400 Dense sphere packings: a blueprint for formal proofs, T. HALES
401 A double Hall algebra approach to affine quantum Schur–Weyl theory, B. DENG, J. DU & Q. FU
402 Mathematical aspects of fluid mechanics, J. ROBINSON, J.L. RODRIGO & W. SADOWSKI

London Mathematical Society Lecture Note Series: 403

Foundations of Computational Mathematics, Budapest 2011

FELIPE CUCKER
City University of Hong Kong

TERESA KRICK
University of Buenos Aires and CONICET

ALLAN PINKUS
Technion

AGNES SZANTO
North Carolina State University

CAMBRIDGE
UNIVERSITY PRESS

University Printing House, Cambridge CB2 8BS, United Kingdom

One Liberty Plaza, 20th Floor, New York, NY 10006, USA

477 Williamstown Road, Port Melbourne, VIC 3207, Australia

314-321, 3rd Floor, Plot 3, Splendor Forum, Jasola District Centre, New Delhi - 110025, India

103 Penang Road, #05-06/07, Visioncrest Commercial, Singapore 238467

Cambridge University Press is part of the University of Cambridge.

It furthers the University's mission by disseminating knowledge in the pursuit of education, learning and research at the highest international levels of excellence.

www.cambridge.org
Information on this title: www.cambridge.org/9781107604070

First published 2013

A catalogue record for this publication is available from the British Library

ISBN 978-1-107-60407-0 Paperback

Contents

Contents

Preface

The Society for the Foundations of Computational Mathematics supports and promotes fundamental research in computational mathematics and its applications, interpreted in the broadest sense. It fosters interaction among mathematics, computer science and other areas of computational science through its conferences, workshops and publications. As part of this endeavour to promote research across a wide spectrum of subjects concerned with computation, the Society brings together leading researchers working in diverse fields. Major conferences of the Society have been held in Park City (1995), Rio de Janeiro (1997), Oxford (1999), Minneapolis (2002), Santander (2005), Hong Kong (2008), and Budapest (2011). The next conference is expected to be held in 2014. More information about FoCM is available at its website http://focm-society.org.

The conference in Budapest on July 4 – 14, 2011, was attended by some 450 scientists. FoCM conferences follow a set pattern: mornings are devoted to plenary talks, while in the afternoon the conference divides into a number of workshops, each devoted to a different theme within the broad theme of foundations of computational mathematics. This structure allows for a very high standard of presentation, while affording endless opportunities for cross-fertilization and communication across subject boundaries. Workshops at the Budapest conference were held in the following nineteen fields:

- Approximation theory
- Asymptotic analysis and high oscillation
- Computational algebraic geometry
- Computational dynamics
- Computational harmonic analysis, image and signal processing
- Computational number theory
- Continuous optimization
- Flocking, swarming, and control of distributed systems
- Foundations of numerical PDEs
- Geometric integration and computational mechanics
- Information-based complexity

- Learning theory
- Multiresolution and adaptivity in numerical PDEs
- Numerical linear algebra
- Random matrix theory, computations & applications
- Real-number complexity
- Special functions and orthogonal polynomials
- Stochastic computation
- Symbolic analysis

In addition to the workshops, eighteen plenary lectures, covering a broad spectrum of topics connected to computational mathematics, were delivered by some of the world's foremost researchers. One of these plenary lectures was presented by Snorre H. Christiansen, as the first recipient of the Stephen Smale Prize awarded by the Society for the Foundations of Computational Mathematics.

This volume is a collection of articles based on the plenary talks presented at FoCM 2011. The topics covered in the lectures and in this volume reflect the breadth of research within computational mathematics as well as the richness and fertility of interactions between seemingly unrelated branches of pure and applied mathematics. The Budapest gathering proved itself to be a stimulating meeting place of researchers in computational mathematics and of theoreticians in mathematics and computer science, with emphasis on multidisciplinary interaction across subjects and disciplines in an informal and friendly atmosphere.

We hope that this volume will be of interest to researchers in the field of computational mathematics and also to non-experts who wish to gain some insight into the state of the art in this active and significant field.

We wish to express our gratitude to the organizing company Scope Meetings Ltd, the Budapest University of Technology and Economics, the host institute, for their technical support, and the local organizing committee for making FoCM 2011 such an outstanding success. We also thank the National Science Foundation (award no. 1068800), and the Commission for Developing Countries of the International Mathematical Union for their financial assistance. We would like to thank the authors of the articles in this volume for producing in short order such excellent contributions. Above all, however, we wish to express our gratitude to all the participants of FoCM 2011 for attending the meeting and making it such an exciting, productive and scientifically stimulating event.

Contributors

Carlos Beltrán *Depto. de Matemáticas, Estadística y Computación, Universidad de Cantabria, Santander, Spain*
beltranc@unican.es

Gunnar Carlsson *Department of Mathematics, Stanford University, Stanford, CA, USA*
gunnar@math.stanford.edu

Snorre H. Christiansen *Centre of Mathematics for Applications & Department of Mathematics, University of Oslo, Oslo, Norway*
s.h.christiansen@cma.uio.no

Steffen Dereich *Fachbereich Mathematik und Informatik, Westfälische Wilhelms-Universität Münster, Münster, Germany*
steffen.dereich@uni-muenster.de

F. Alberto Grünbaum *Department of Mathematics, University of California, Berkeley, CA, USA*
grunbaum@math.berkeley.edu

Ernst Hairer *Section de Mathématiques, Université de Genève, Genève, Switzerland*
Ernst.Hairer@unige.ch

Thomas Y. Hou *Applied and Comput. Math, Caltech, Pasadena, CA, USA*
hou@cms.caltech.edu

Evelyne Hubert *INRIA Méditérranée, Sophia Antipolis, France*
Evelyne.Hubert@inria.fr

Christian Lubich *Mathematisches Institut, Universität Tübingen, Tübingen, Germany*
Lubich@na.uni-tuebingen.de

Thomas Müller-Gronbach *Fakultät für Informatik und Mathematik, Universität Passau, Passau, Germany*
thomas.mueller-gronbach@uni-passau.de

Hans Z. Munthe-Kaas *Department of Mathematics, University of Bergen, Bergen, Norway*
hans.munthe-kaas@math.uib.no

Morten Nome *Department of Mathematics, University of Bergen, Bergen, Norway*
morten.nome@math.uib.no

Klaus Ritter *Fachbereich Mathematik, Technische Universität Kaiserslautern, Kaiserslautern, Germany*
ritter@mathematik.uni-kl.de

Brett N. Ryland *Department of Mathematics, University of Bergen, Bergen, Norway*
nappers@gmail.com

Zuoqiang Shi *Mathematical Sciences Center, Tsinghua University, Beijing, China*
zqshi@math.tsinghua.edu.cn

William Stein *Department of Mathematics, University of Washington, Seattle, WA, USA*
wstein@gmail.com

Luis Velázquez *Departamento de Matemática Aplicada, Universidad de Zaragoza, Zaragoza, Spain*
velazque@unizar.es

Shu Wang *College of Applied Sciences, Beijing University of Technology, Beijing, China*
wangshu@bjut.edu.cn

1

The State of the Art in Smale's 7th Problem

Carlos Beltrán[a]
Depto. de Matemáticas, Estadística y Computación
Universidad de Cantabria

1.1 A very brief historical note

Smale's 7th problem is the computational version of an old problem dating back to Thomson [30] and Tammes [29], see Whyte's early review [32] for its history, namely, the sensible distribution of points in the two–dimensional sphere. In Whyte's paper different possible definitions of "well–distributed points in the sphere" are suggested:

1. Points which maximise the product of their mutual distances (called elliptic Fekete points [1] after [14]).
2. Points which minimise the sum of the inverse of their mutual distances (Thomson's problem), and more generally which minimise some sum of potentials which depend on the mutual distances (like Riesz potentials).
3. Points which maximise the least distance between any pair.
4. Points which are the center of the optimal packing problem, that is, the problem of finding the smallest radius of a sphere such that one can place on its surface k non–overlapping circles of a given radius.

This beautiful problem is terribly challenging! A first shocking result by Leech [19] showed that even though the set of N particles on the sphere which are critical points for the problem in item (2) for *every possible* potential can be completely described, this description is not enough to solve the problem for *any particular* potential. Namely, solving problem (2) for some particular potential may be completely meaningless for solving problem (2) for another, different potential. We quote Leech:

[a] Partially supported by MTM2010-16051 (Spanish Ministry of Science and Innovation MICINN).

[1] Not to be confused with the so called Fekete points.

There is no obvious way of relating the present problems to other extremal problems such as minimising the greatest distance at which an arbitrary point can be placed from the nearest point of a configuration. In fact, since a configuration which is not balanced is out of equilibrium under almost all laws of force, it is not to be expected that any such configuration will be found to be of significance in respect to both an equilibrium problem and another extremal problem, or even under two different significant equilibrium problems.

The problem has so many ramifications that it is difficult even to mention all of them. There are dozens of papers written about each of the mentioned problems. In this paper we focus only on the explicit version proposed by Smale [27]: the problem of finding elliptic Fekete points in the two–dimensional sphere. Our reference list will also refer only to this problem, and thus many important articles dealing with the other versions are omitted, for the sake of brevity.

1.2 The problem

Given N different points $x_1, \ldots, x_N \in \mathbb{R}^3$, let $X = (x_1, \ldots, x_N)$ and

$$\mathcal{E}(X) = \mathcal{E}(x_1, \ldots, x_N) = -\sum_{i<j} \log \|x_i - x_j\|$$

be its *logarithmic potential* or *logarithmic energy* [2]. Let

$$\mathbb{S} = \{(a,b,c) \in \mathbb{R}^3 : a^2 + b^2 + (c - 1/2)^2 = 1/4\}$$

$$= \{(a,b,c) \in \mathbb{R}^3 : a^2 + b^2 + c^2 = c\}$$

be the Riemann sphere, i.e. the sphere in $\mathbb{R}^3$ of radius $1/2$ centered at $(0,0,1/2)^T$, and let

$$m_N = \min_{x_1,\ldots,x_N \in \mathbb{S}} \mathcal{E}(x_1, \ldots, x_N)$$

be the minimum value of $\mathcal{E}$. A minimising N–tuple $X = (x_1, \ldots, x_N)$ is called a set of elliptic Fekete points. Note that such a N–tuple can also be defined as a set of N points in the sphere which *maximise the product of their mutual distances.*

Smale's 7th problem [27]: *Can one find $X = (x_1, \ldots, x_N)$ such that*

$$\mathcal{E}(X) - m_N \leq c \log N, \qquad c \text{ a universal constant.} \tag{1.1}$$

[2] Sometimes $\mathcal{E}(X)$ is denoted by $\mathcal{E}_0(X)$, $\mathcal{E}(0, X)$ or $V_N(X)$

By "can one find" Smale means "can one describe an algorithm (in the BSS model of computation[3]) which on input N produces such a N–tuple, in running time bounded by a polynomial in N?".

Remark 1.1 The original question in [27] is not for points in $\mathbb{S}$ but for points in the unit sphere $\{(a, b, c) \in \mathbb{R}^3 : a^2 + b^2 + c^2 = 1\}$. We prefer to use the Riemann sphere instead of the unit sphere because some of the results look more natural when stated in $\mathbb{S}$. Another powerful reason to do so is that $-\log \|x - y\|$ is positive for every $x, y \in \mathbb{S}$ while the same claim is not true for x, y in the unit sphere. This helps intuition at some moments. Of course, the problems of finding a set of elliptic Fekete points in $\mathbb{S}$ or in the unit sphere are equivalent via the transformation

$$(a, b, c) \in \mathbb{S} \mapsto (2a, 2b, 2c - 1).$$

If $x_1, \ldots, x_N \in \mathbb{S}$ and we denote $\hat{x}_1, \ldots, \hat{x}_N$ their associated unit sphere points via this transform, then we have:

$$\mathcal{E}(\hat{x}_1, \ldots, \hat{x}_N) = \mathcal{E}(x_1, \ldots, x_N) - \frac{\log 2}{2} N(N-1)$$

We thus state all the results for $\mathbb{S}$, translating them from their original citations when necessary.

1.3 The value of m_N

The first problem one encounters in dealing with Smale's 7th problem is that the value of m_N is not known, even to $O(N)$. A general technique (valid for Riemannian manifolds) given by Elkies shows that

$$m_N \geq \frac{N^2}{4} - \frac{N \log N}{4} + O(N).$$

Wagner [31] used the stereographic projection and Hadamard's inequality to get another lower bound. His method was refined by Rakhmanov, Saff and Zhou [21], who also proved an upper bound for m_N using partitions of the sphere. The lower bound was subsequently improved upon

[3] For the nonexpert reader, a BSS algorithm is just an algorithm in the natural sense of the word: a sequence of instructions (arithmetic operations, comparisons and, in general, any of the usual instructions present in a computer program) that, correctly executed, gives an answer. The arithmetic operations are assumed to be exact when performed on real numbers. See [9, 8] for details.

by Dubickas and Brauchart [13], [10]. The following result summarizes the best known bounds:

Theorem 1.2 *Let C_N be defined by*

$$m_N = \frac{N^2}{4} - \frac{N \log N}{4} + C_N N.$$

Then,

$$-0.4375 \leq \liminf_{N \mapsto \infty} C_N \leq \limsup_{N \mapsto \infty} C_N \leq -0.3700708...$$

1.4 The separation distance

The separation distance of a N–tuple $X = (x_1, \ldots, x_N) \in \mathbb{S}^N$ is defined by

$$d_{\|\cdot\|,sep}(X) = \min_{i \neq j} \|x_i - x_j\|.$$

By the definition of $\mathcal{E}$, it is clear that if X is a set of elliptic Fekete points then $d_{\|\cdot\|,sep}(X)$ cannot be too small. Using tools from classical potential theory, Rakhmanov, Saff and Zhou [22, 21] first proved the lower bound $3/(10\sqrt{N})$ for the separation distance of a set of elliptic Fekete points. Their result was improved by Dubickas [13] to $7/(8\sqrt{N})$. The sharpest known bound is due to Dragnev [11]:

Theorem 1.3 *Let X be a set of elliptic Fekete points. Then,*

$$d_{\|\cdot\|,sep}(X) \geq \frac{1}{\sqrt{N-1}}.$$

Recall that given two points $x, y \in \mathbb{S}$, the Riemannian distance $d_R(x, y)$ is the length of the shortest curve in $\mathbb{S}$ joining x and y. Elementary trigonometry shows that

$$d_R(x, y) = \arcsin \|x - y\|.$$

Thus, if we define

$$d_{R,sep}(X) = \min_{i \neq j} d_R(x_i, x_j),$$

we have

$$d_{R,sep}(X) \geq \arcsin \frac{1}{\sqrt{N-1}} \geq \frac{1}{\sqrt{N-1}},$$

for X as above.

1.5 The condition number of polynomials and Bombieri–Weyl norm

According to [27], one of Smale's motivations for studying the problem of elliptic Fekete points was to find polynomials all of whose zeros are well conditioned. Shub and Smale [24] defined a certain quantity (the "condition number") and used it to measure the stability and complexity of polynomial zero finding algorithms. Given a homogeneous polynomial

$$h(z,t) = \sum_{k=0}^{N} a_k z^k t^{N-k}, \qquad a_k \in \mathbb{C}, a_N \neq 0$$

of degree $N \geq 1$, and a projective zero $\zeta \in \mathbb{P}(\mathbb{C}^2)$ of h, the condition number of h at ζ is[4]

$$\mu(h,\zeta) = N^{1/2} \|(Dh(\zeta)\mid_{\zeta^\perp})^{-1}\| \|h\| \|\zeta\|^{N-1},$$

or $+\infty$ if $Dh(\zeta)\mid_{\zeta^\perp}$ is not an invertible mapping. Here, $Dh(\zeta)\mid_{\zeta^\perp}$ is the restriction of the derivative to the orthogonal complement of ζ in $\mathbb{C}^2$, and

$$\|h\| = \left(\sum_{k=0}^{N} \binom{N}{k}^{-1} |a_k|^2 \right)^{1/2}$$

is the Bombieri–Weyl norm (sometimes called the Kostlan norm) of h. If no zero of h is specified, we just take the maximum:

$$\mu(h) = \max_{\zeta \in \mathbf{P}(\mathbb{C}^2): h(\zeta)=0} \mu(h,\zeta).$$

Now, let f be a univariate polynomial

$$f(X) = \sum_{k=0}^{N} a_k X^k, \qquad a_k \in \mathbb{C}, a_N \neq 0,$$

and let $z \in \mathbb{C}$ be a zero of f. We define

$$\mu(f,z) = \mu(h,(z,1)), \qquad \mu(f) = \max_{z \in \mathbb{C}: f(z)=0} \mu(f,z),$$

where $h(X,Y) = \sum_{k=0}^{N} a_k X^k Y^{N-k}$ is the homogeneous counterpart of f. Taking $\|f\| = \|h\|$, one can write:

$$\mu(f,z) = \frac{N^{1/2}(1+|z|^2)^{\frac{N-2}{2}}}{|f'(z)|} \|f\|.$$

[4] Sometimes μ is denoted μ_{norm} or μ_{proj} but we here keep the simpler notation.

In [26] Shub and Smale proved the following relation between the condition number and elliptic Fekete points. Let $\mathfrak{Re}$ and $\mathfrak{Im}$ be, respectively, the real and complex part of a complex number.

Theorem 1.4 *Let $z_1, \ldots, z_N \in \mathbb{C}$ be a set of complex numbers. For $1 \le i \le N$, let $x_i \in \mathbb{S}$ be the preimage of z_i under the stereographic projection, that is*

$$x_i = \left(\frac{\mathfrak{Re}(z_i)}{1+|z_i|^2}, \frac{\mathfrak{Im}(z_i)}{1+|z_i|^2}, \frac{1}{1+|z_i|^2} \right)^T \in \mathbb{S}, \qquad 1 \le i \le N. \tag{1.2}$$

Assume that $x_1, \ldots, x_N$ are a set of elliptic Fekete points. Let $f : \mathbb{C} \to \mathbb{C}$ be a degree N polynomial such that its zeros are $z_1, \ldots, z_N$. Then,

$$\mu(f) \le \sqrt{N(N+1)}.$$

More generally, let $z_1, \ldots, z_N \in \mathbb{C}$ be any collection of N distinct complex numbers, let f be a polynomial with zeros $z_1, \ldots, z_N$ and let $x_1, \ldots, x_N$ be given by (1.2)*. Then,*

$$\mu(f) \le \sqrt{N(N+1)} \frac{e^{\mathcal{E}(x_1, \ldots, x_N)}}{e^{m_N}}.$$

It is interesting to remark that there exists no explicit known way of describing a sequence of polynomials satisfying $\mu(f) \le N^c$, for any fixed constant c and $N \ge 1$. Theorem 1.4 says that, if a N–tuple satisfying (1.1) can be described for any N, then such a sequence of polynomials can also be generated.

Here is a nice formula (which just follows from the definitions) relating $\mathcal{E}$ to μ and Bombieri–Weyl norm:

$$\mathcal{E}(x_1, \ldots, x_N) = \frac{1}{2} \sum_{i=1}^N \log \mu(f, z_i) + \frac{N}{2} \log \frac{\prod_{i=1}^N \sqrt{1+|z_i|^2}}{\|f\|} - \frac{N}{4} \log N.$$

Note that the term

$$\frac{\prod_{i=1}^N \sqrt{1+|z_i|^2}}{\|f\|} \tag{1.3}$$

in the previous formula is the quotient between the product of the Bombieri–Weyl norm of the factors of f and the Bombieri–Weyl norm of f. That quantity is always greater than 1, see [3]. Experiments suggest that minimising $\mathcal{E}$ is a problem similar to minimising the sum of $\log \mu(f, z_i)$, and to maximising the quotient (1.3)[5]. We recall from [3,

[5] This may seem surprising at a first glance. It turns out that (1.3) is minimal, i.e., equal to 1, precisely when all the z_i are equal, which implies $\mathcal{E}(x_1, \ldots, x_N) = \infty$.

Theorem 2.1] (see also [4]) that for two polynomials f, g of respective degrees r and s,

$$\|f \cdot g\| \geq \sqrt{\frac{r!s!}{(r+s)!}} \|f\| \cdot \|g\|, \tag{1.4}$$

and this bound is optimal. Maximising (1.3) would be solved by finding an analogue of (1.4) for products of N polynomials, a nice mathematical problem in its own right. As pointed out in [4], it follows from (1.4) that

$$\frac{\prod_{i=1}^{N} \sqrt{1+|z_i|^2}}{\|f\|} \leq \sqrt{N!},$$

but this inequality is far from optimal (for example, if the $z_i's$ are the Nth roots of unity, then the value of this quotient is $\sqrt{2^{N-1}} \ll \sqrt{N!}$).

1.6 The average value and random polynomials

Random polynomials[6] have been known, since [25], to be well–conditioned on average, meaning that their condition number is polynomially bounded by their degree, on average. This, combined with Theorem 1.4, suggests that spherical points associated with zeros of random polynomials should produce small values of $\mathcal{E}$. To properly state this fact, let us consider $\mathcal{E}$ as a function defined on $\mathbb{S}^N \setminus \Sigma$ where

$$\Sigma = \{(x_1, \ldots, x_N) \in \mathbb{S}^N : x_i = x_j \text{ for some } i \neq j\}.$$

Note that Σ is the set of N–tuples $(x_1, \ldots, x_N)$ of points in $\mathbb{S}$ such that the polynomial f whose zeros are associated with the x_i satisfies $\mu(f) = \infty$.

Here and throughout this paper, for a measure space X, a measurable, finite volume subset $U \subseteq X$, and a measurable function $f : U \to [0, \infty)$ we set

$$\fint_U f = \frac{\int_U f}{\int_U 1} = \frac{1}{\text{Volume}(U)} \int_U f.$$

One can easily compute the average value of $\mathcal{E}$ when $x_1, \ldots, x_N$ are chosen at random in $\mathbb{S}$, uniformly and independently with respect to the probability distribution induced by Lebesgue measure in $\mathbb{S}$:

$$\fint_{x \in \mathbb{S}^N \setminus \Sigma} \mathcal{E}(X) = \frac{N^2}{4} - \frac{N}{4}.$$

[6] in the sense of Theorem 1.5 below.

By comparing this with Theorem 1.2, we can see that random choices of points in the sphere already produce pretty low values of the minimal energy. One can ask if other (simple) probability distributions produce even lower average results. In [2] we proved a relation with random polynomials.

Theorem 1.5 *Let $f(X) = \sum_{k=0}^{N} a_k X^k$ be a random polynomial where the a_k are independent complex random variables, such that the real and imaginary parts of a_k are independent (real) Gaussian random variables centered at 0 with variance $\binom{N}{k}$. Let $z_1, \ldots, z_N$ be the complex zeros of f, and let x_i be given by (1.2). Then, the expected value of $\mathcal{E}(x_1, \ldots, x_N)$ equals*

$$\frac{N^2}{4} - \frac{N \log N}{4} - \frac{N}{4}.$$

Again, by comparing this with Theorem 1.2, we conclude that spherical points coming from zeros of random polynomials are pretty well distributed, as they agree with the minimal value of $\mathcal{E}$, to order $O(N)$. This result fits into a more general (yet, less precise) kind of result related to random sections on Riemann surfaces, see [33, 34]. Note that the notion of "random polynomial" used in Theorem 1.5 is strongly related to the Bombieri–Weyl norm. It is the natural Gaussian distribution associated with the space of polynomials, considered as a normed vector space with the Bombieri–Weyl norm.

1.7 Properties of the critical points of $\mathcal{E}$

One of the first things to do when faced with an optimization problem is to study the critical points of the objective function, i.e. the points where the derivative vanishes. In our case the derivative of $\mathcal{E}$ is easy to compute. Algebraic manipulation of its expression was used in [7, 12] to get the following [7]:

Theorem 1.6 *Let $x_1, \ldots, x_N \in \mathbb{S}$ be a critical point of $\mathcal{E}$. Let*

$$\mathfrak{o} = \left(0, 0, \frac{1}{2}\right)^T$$

be the center of $\mathbb{S}$. Then,

[7] In [12] the result is stated for global minima, but the proof is indeed valid for any critical point of $\mathcal{E}$. Moreover, the third item in this theorem is here stated in greater generality than in [12], but the same proof holds.

- *The center of mass of the x_i is $\mathfrak{o}$. Namely,*

$$\sum_{i=1}^{N} \overrightarrow{\mathfrak{o}x_i} = \mathfrak{o}, \qquad \text{or equivalently} \qquad \frac{1}{N}\sum_{i=1}^{N} x_i = \mathfrak{o}.$$

- *For every $1 \leq i \leq N$, we have:*

$$\sum_{j\neq i} \frac{\overrightarrow{x_j x_i}}{\|x_j - x_i\|^2} = 2(N-1)\overrightarrow{\mathfrak{o}x_i}.$$

- *For every $x \in \mathbb{S}$, we have:*

$$\sum_{i=1}^{N} \|x - x_i\|^2 = \frac{N}{2}.$$

Other natural questions are, which types of critical points does $\mathcal{E}$ have and how many of them exist? There are some conjectures about their number (some authors conjecture that the number of local minima grows exponentially on N, see the references in Section 1.12 below) but no precise result is known. It was pointed out in [26] that there exist critical points of $\mathcal{E}$ of index N, namely N points evenly distributed on some equator of $\mathbb{S}$. It follows from (1.5) below and the maximum principle of harmonic analysis that no local maximum of $\mathcal{E}$ can exist.

1.8 Harmonic properties of $\mathcal{E}$

Let us endow $\mathbb{S}^N$ with its natural Riemannian structure, that is the product structure (or equivalently, the structure inherited from $\mathbb{R}^{3N}$). Again viewing $\mathcal{E}$ as a function $\mathcal{E} : \mathbb{S}^N \setminus \Sigma \to \mathbb{R}$, we computed in [5] its (Riemannian) Laplacian. It turns out that

$$\Delta\mathcal{E} \equiv 2N(N-1), \tag{1.5}$$

is a constant. If a function defined on an open set of $\mathbb{R}^n$ has a constant Laplacian, then the classical mean value theorem of harmonic analysis gives a formula for the mean value of the function on a ball centered at every point. In the case of $\mathbb{S}^N$, one can use the theory of harmonic manifolds to analyze the mean value of $\mathcal{E}$ in products of spherical caps, that is in sets of the form

$$\begin{aligned} B_\infty(X, \vec{\varepsilon}) &= \{(y_1, \ldots, y_N) \in \mathbb{S}^N : d_R(x_i, y_i) < \varepsilon_i, 1 \leq i \leq N\} \\ &= B(x_1, \varepsilon_1) \times \cdots \times B(x_N, \varepsilon_N) \subseteq \mathbb{S}^N, \end{aligned}$$

where $X = (x_1, \ldots, x_N)$, $\vec{\varepsilon} = (\varepsilon_1, \ldots, \varepsilon_N)$ and for $x \in \mathbb{S}$, $\varepsilon > 0$, $B(x, \varepsilon)$ is the open spherical cap of (Riemannian) radius equal to ε. Abusing notation, if $\varepsilon = 0$ we define $B(x, 0) = \{x\}$. The mean value of $\mathcal{E}$ in $B_\infty(X, \vec{\varepsilon})$ was studied in [5]:

Theorem 1.7 *Let $X \in \mathbb{S}^N \setminus \Sigma$ and $\vec{\varepsilon} \in [0, \pi/2)^N$ be such that $B_\infty(X, \vec{\varepsilon}) \subseteq \mathbb{S}^N \setminus \Sigma$. Then,*

$$\fint_{B_\infty(X,\vec{\varepsilon})} \mathcal{E}(Y)\, dY = \mathcal{E}(X) + C_N(\vec{\varepsilon}),$$

where

$$C_N(\vec{\varepsilon}) = (N-1) \sum_{j=1}^{N} \left(\frac{1}{2} + \frac{\log(\cos \varepsilon_j)}{\tan^2 \varepsilon_j} \right) \in \left[0, \frac{N-1}{2}\right),$$

with the convention that

$$\frac{1}{2} + \frac{\log(\cos 0)}{\tan^2 0} = 0.$$

The reader may find useful the estimate $\frac{1}{2} + \frac{\log(\cos \varepsilon)}{\tan^2 \varepsilon} \approx \frac{\varepsilon^2}{4}$ for small values of ε.

1.9 The limiting distribution

It follows from classical potential theory that optimal logarithmic energy points are uniformly distributed over $\mathbb{S}$, asymptotically as $N \mapsto \infty$, in the following sense: Let $\{X^{(N)} = (x_1^{(N)}, \ldots, x_N^{(N)})\}$ be a sequence such that $X^{(N)} \in \mathbb{S}^N$ is a set of N elliptic Fekete points in $\mathbb{S}$ for every $N \geq 2$. Then, for any continuous function $f : \mathbb{S} \to \mathbb{R}$ we have:

$$\fint_{\mathbb{S}} f = \lim_{N \mapsto \infty} \frac{1}{N} \sum_{j=1}^{N} f(x_j^{(N)}). \tag{1.6}$$

One way to analyze this qualitative result is to study the so called spherical cap discrepancy, that is for fixed $N \geq 2$:

$$D_C(X^{(N)}) = \sup_{C} \left| \frac{\sharp(X^{(N)} \cap C)}{N} - \fint_{\mathbb{S}} \chi_C \right|,$$

where χ_C is the characteristic function of C and the supremum is taken over all possible spherical caps C in $\mathbb{S}$. Note that $D_C(X^{(N)})$ measures how far the counting measure is from the probability measure associated with Lebesgue measure in $\mathbb{S}$. In [10], Brauchart proved the following estimate:

Theorem 1.8 *Let* $\{X^{(N)} = (x_1^{(N)}, \ldots, x_N^{(N)})\}$ *be a sequence such that* $X^{(N)} \in \mathbb{S}^N$ *is a set of elliptic Fekete points for every* $N \geq 2$*. Then,*

$$D_C(X^{(N)}) \leq cN^{-1/4},$$

for some constant c*.*

We also note the following interesting corollary from [10] [8]:

Corollary 1.9 *In the notation of Theorem 1.8, let* $f : \mathbb{S} \to \mathbb{R}$ *be a Lipschitz continuous function with Lipschitz modulus* K*. There then exists a constant* $C \geq 0$ *such that*

$$\left| \fint_{\mathbb{S}} f - \frac{1}{N} \sum_{j=1}^{N} f(x_j^{(N)}) \right| \leq K(C + \|f\|_\infty) N^{-1/4}.$$

This gives a more detailed version of (1.6).

1.10 Admissible errors and an exponential time algorithm

A question we can ask about Smale's 7th problem is, how close must one be to an actual set of elliptic Fekete points to ensure that (1.1) is satisfied? One can analyze this question using the *admissible error function* $\mathbf{e} : [0, \infty) \to [0, \infty)$ given by:

$$\mathbf{e}(t) = \sup\{\varepsilon : Y \in B_\infty(X, \vec{\varepsilon}) \text{ implies } \mathcal{E}(Y) \leq m_N + t\},$$

where $X = (x_1, \ldots, x_N)$ is a set of elliptic Fekete points and $\vec{\varepsilon} = (\varepsilon, \ldots, \varepsilon)$. Note that the supremum is indeed a maximum and that, for fixed $t \in [0, \infty)$, $\mathbf{e}(t)$ is the maximum coordinate–wise error (measured in Riemannian distance in $\mathbb{S}$) that one can permit when writing a set of elliptic Fekete points, if an inequality $\mathcal{E}(Y) \leq m_N + t$ is to be guaranteed. Using Theorem 1.7, the following result was proved in [5]:

Theorem 1.10 *Let* $N \geq 3$ *and let* $d_{R,sep}$ *be the Riemannian separation distance of some set of elliptic Fekete points. The admissible error function satisfies*[9]

$$\mathbf{e}(t) \in \left[\sqrt{\frac{t}{2N^2(N-1)}}, \sqrt{\frac{2t}{N(N-1)}} \right], \qquad \textit{for } 0 \leq t \leq \frac{1}{18}.$$

[8] The version in [10] is more precise, we just include a simplified version here.
[9] Again, we just include a simplified version of this result.

In particular, we have

$$\mathbf{e}\left(\frac{1}{18}\right) \in \left[\frac{1/6}{N\sqrt{N-1}}, \frac{1/3}{\sqrt{N(N-1)}}\right]. \quad (1.7)$$

The meaning of (1.7) is the following. If a N–tuple Y satisfying $\mathcal{E}(Y) \leq m_N + 1/18$ is desired, then we can make an error of approximately $1/(6N^{3/2})$ in the description of each coordinate, but if we make an error greater than approximately $1/(3N)$ then we risk that our N–tuple will not satisfy the desired inequality.

The set of rational points (i.e. points with rational coordinates) is dense in $\mathbb{S}$. Moreover, precise bounds on the size of the rational numbers present in a spherical cap of a given radius are known, see for example [23]. A consequence is the following. There exists a N–tuple $X = (x_1, \ldots, x_N)$ of points in $\mathbb{S}$ such that $\mathcal{E}(X) \leq m_N + 1/18$ and such that every x_i has rational coordinates which can be written using integers of absolute value at most $(cN)^6$, $c \leq 17$, a constant. Thus, Smale's 7th problem can be transformed into a discrete search problem: among all such N–tuples, find the one which has smallest value of $\mathcal{E}$. Simply by testing all of them, we get:

Theorem 1.11 *There is an exponential time algorithm (on the BSS model and also on the Turing machine model) for Smale's 7th problem.*

To the knowledge of the author, this algorithm is the only one which has been proved to find N–tuples satisfying (1.1). Of course, its exponential running time makes it impractical.

1.11 Particular values of N

For some values of N, sets of elliptic Fekete points have been encountered:

- For $N = 2$, two antipodal points.
- For $N = 3$, the vertices of an equilateral triangle lying on any equator of $\mathbb{S}$.
- For $N = 4$, the vertices of a regular tetrahedron (see [12, Cor. 3] for a more general version of this fact).
- For $N = 5$, two antipodal points, say at the North and South Pole, and three that form an equilateral triangle on the Equator (see [12, Th. 1]).

- For $N = 6$, the vertices of a regular octahedron (see [17])
- For $N = 12$, the vertices of the regular icosahedron (see [1]).

Among these, the cases $N = 2, 3$ are trivial and the case $N = 4$ is easy. The rest of the cases have more difficult proofs. Note that, for $N = 5, 6, 12$, it is already nontrivial to prove that there is a unique (up to rotations and symmetries) set of elliptic Fekete points. Another remark is that the vertices of the platonic bodies do not always form a set of elliptic Fekete points: the value of $\mathcal{E}$ at the vertices of a regular cube inscribed in $\mathbb{S}$ is greater than the value at the same set of vertices, when four of them sharing a plane are rotated 45 degrees.

1.12 Numerical experiments

Numerical computation of elliptic Fekete points is a hard problem with "massive multiextremality" (see [16]): it is known that $\mathcal{E}$ has many saddle points, and numerical analysts believe that there exist many local minima, its number possibly growing exponentially with N, some (or all) of them having values of $\mathcal{E}$ very close to the (conjectured) global minimal value. Due to this, and to the increasing amount of resources needed to numerically determine solutions, Smale's 7th problem is well–suited for testing global optimization routines. There have been many attempts which have produced detailed formulas, inspiring pictures and thrilling conjectures. We do not intend to reproduce those works here, but we encourage the reader to read them in [7, 21, 35, 22, 18, 28, 20, 15, 6] and references therein.

References

[1] N. N. Andreev, An extremal property of the icosahedron. *East J. Approx.*, **2**, 459–462, 1996.

[2] D. Armentano, C. Beltrán, and M. Shub, Minimizing the discrete logarithmic energy on the sphere: The role of random polynomials. *Trans. Amer. Math. Soc.*, **363**, 2955–2965, 2011.

[3] B. Beauzamy, E. Bombieri, P. Enflo, and H. L. Montgomery, Products of polynomials in many variables. *J. Number Theory*, **36**, 219–245, 1990.

[4] B. Beauzamy, V. Trevisan, and P. S. Wang, Polynomial factorization: sharp bounds, efficient algorithms. *J. Symbolic Comput.*, **15**, 393–413, 1993.

[5] C. Beltrán, Harmonic properties of the logarithmic potential and the computability of elliptic Fekete points. To appear in *Constr. Approx.* (DOI: 10.1007/s00365-012-9158-y).

[6] E. Bendito, A. Carmona, A. M. Encinas, J. M. Gesto, A. Gómez, C. Mouriño, and M. T. Sánchez, Computational cost of the Fekete problem. I. The forces method on the 2-sphere. *J. Comput. Phys.*, **228**, 3288–3306, 2009.

[7] B. Bergersen, D. Boal, and P. Palffy-Muhoray, Equilibrium configurations of particles on a sphere: the case of logarithmic interactions. *J. Phys. A: Math. Gen.*, **27**, 2579–2586, 1994.

[8] L. Blum, F. Cucker, M. Shub, and S. Smale, *Complexity and Real Computation.* Springer-Verlag, New York, 1998.

[9] L. Blum, M. Shub, and S. Smale, On a theory of computation and complexity over the real numbers: NP-completeness, recursive functions and universal machines. *Bull. Amer. Math. Soc. (N.S.)*, **21**, 1–46, 1989.

[10] J. S. Brauchart, Optimal logarithmic energy points on the unit sphere. *Math. Comp.*, **77**, 1599–1613, 2008.

[11] P. D. Dragnev, On the separation of logarithmic points on the sphere. In *Approximation Theory X, (St. Louis, MO, 2001)*, Innov. Appl. Math., p. 137–144, Vanderbilt Univ. Press, Nashville, TN, 2002.

[12] P. D. Dragnev, D. A. Legg, and D. W. Townsend, Discrete logarithmic energy on the sphere. *Pacific J. Math.*, **207**, 345–358, 2002.

[13] A. Dubickas, On the maximal product of distances between points on a sphere. *Liet. Mat. Rink.*, **36**, 303–312, 1996.

[14] M. Fekete, Über die Verteilung der Wurzeln bei gewissen algebraischen Gleichungen mit ganzzahligen Koeffizienten. *Math. Z.*, **17**, 228–249, 1923.

[15] D. Hardin and E. B. Saff, Discretizing manifolds via minimum energy points. *Notices Amer. Math. Soc.*, **51**, 1186–1194, 2004.

[16] J.-B. Hiriart-Urruty, A new series of conjectures and open questions in optimization and matrix analysis. In *ESAIM : Control, Optimisation and Calculus of Variations*, p. 454–470, 2009.

[17] A. V. Kolushov and V. A. Yudin, Extremal dispositions of points on the sphere. *Anal. Math.*, **23**, 25–34, 1997.

[18] A. B. J. Kuijlaars and E. B. Saff, Distributing many points on a Sphere. *Math. Int.*, **19**, 5–11, 1997.

[19] J. Leech, Equilibrium of sets of particles on a sphere. *Math. Gaz.*, **41**, 81–90, 1957.

[20] J. Pintér, Globally optimized spherical point arrangements: Model variants and illustrative results. *Ann. Op. Res.*, **104**, 213–230, 2001.

[21] E. A. Rakhmanov, E. B. Saff, and Y. M. Zhou, Minimal discrete energy on the sphere. *Math. Res. Letters*, **1**, 647–662, 1994.

[22] E. A. Rakhmanov, E. B. Saff, and Y. M. Zhou, Electrons on the sphere. In *Computational Methods and Function Theory 1994 (Penang)*, volume 5 of *Ser. Approx. Decompos.*, p. 293–309, World Sci. Publ., River Edge, NJ, 1995.

[23] E. Schmutz, Rational points on the unit sphere. *Cent. Eur. J. Math.*, **6**, 482–487, 2008.

[24] M. Shub and S. Smale, Complexity of Bézout's theorem. I. Geometric aspects. *J. Amer. Math. Soc.*, **6**, 459–501, 1993.

[25] M. Shub and S. Smale, Complexity of Bezout's theorem. II. Volumes and probabilities. In *Computational Algebraic Geometry (Nice, 1992)*, volume 109 of *Progr. Math.*, p. 267–285, Birkhäuser Boston, Boston, MA, 1993.

[26] M. Shub and S. Smale, Complexity of Bezout's theorem. III. Condition number and packing. *J. Complexity*, **9**, 4–14, 1993. Festschrift for Joseph F. Traub, Part I.

[27] S. Smale, Mathematical problems for the next century. In *Mathematics: Frontiers and Perspectives*, p. 271–294. Amer. Math. Soc., Providence, RI, 2000.

[28] W. Stortelder, J. Swart, and J. Pintér, Finding elliptic Fekete points sets: two numerical solution approaches. *J. Comp. App. Math.*, **130**, 205-216, 2001.

[29] P. M. L. Tammes, *On the origin of number and arrangement of the places of exit on the surface of pollen-grains.* Recueil des travaux botaniques neerlandais 27, Diss. Groningen., 1930.

[30] J. J. Thomson, On the structure of the atom. *Phil. Mag.*, **7**, 237–265, 1904.

[31] G. Wagner, On the product of distances to a point set on a sphere. *J. Austral. Math. Soc. Ser. A*, **47**, 466–482, 1989.

[32] L. L. Whyte, Unique arrangements of points on a sphere. *Amer. Math. Monthly*, **59**, 606–611, 1952.

[33] S. Zelditch and Q. Zhong, Addendum to "Energies of zeros of random sections on Riemann surfaces". *Indiana Univ. Math. J.*, **59**, 2001–2006, 2010.

[34] Q. Zhong, Energies of zeros of random sections on Riemann surfaces. *Indiana Univ. Math. J.*, **57**, 1753–1780, 2008.

[35] Y. Zhou, *Arrangements of Points on the Sphere.* Ph. D. Thesis, Math. Department, University of South Florida, 1995.

2

The Shape of Data

Gunnar Carlsson[a]
Department of Mathematics
Stanford University

2.1 Introduction

Data is often analyzed by approximating it via models of various types. For example, regression methods are often used to model the data as the graph of a function, in one or more variables. Often, though, certain qualitative properties which one can readily observe when the data is two-dimensional are of a great deal of importance for understanding it, and these features are not readily represented within this model.

Example 2.1 This data below appears to be divided into three disconnected groups.

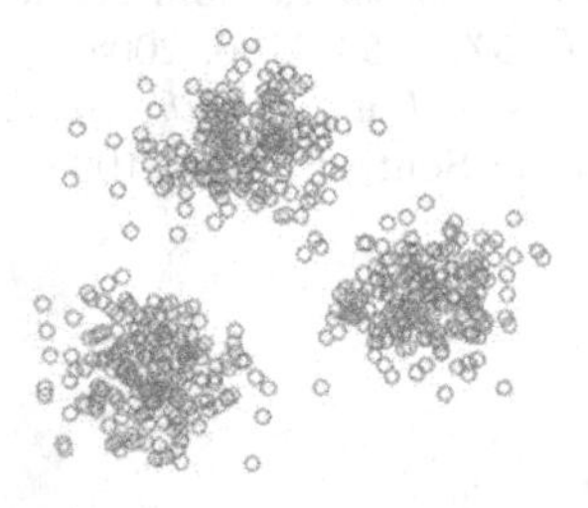

For example, it might be data about various physical characteristics coming from three different population groups, or biomedical data coming from different forms of a disease. Seeing that the data breaks into groups in this fashion can give insight into the data, once one understands what characterizes them.

[a] Supported in part by NIH Grant I-U54 CA149145-01, Air Force Office of Scientific Research Grant FA9550-09-0-1-0531, Office of Naval Research Grant N00014-08-1-0931, and NSF DMS 0905823.

Example 2.2 This data set is obtained from a Lotka-Volterra equation modeling the populations of predators and preys over time.

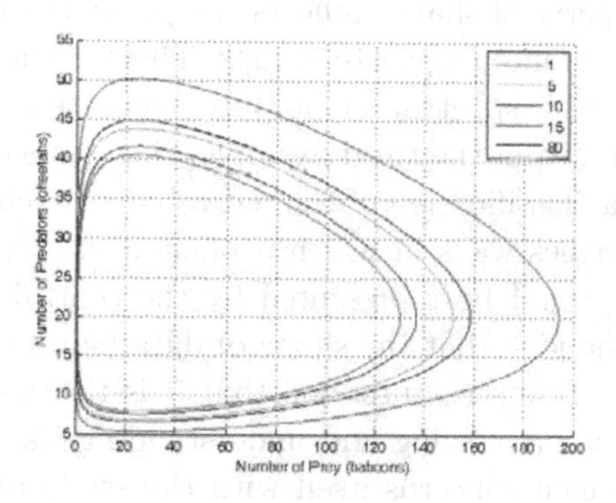

The first observation about this data is that it is arranged in a loop. The loop is not exactly circular, but it is certainly topologically a circle. The exact form of the equations, while interesting, is not of as much importance as this qualitative observation, which reflects the fact that the underlying phenomenon is recurrent or periodic. If we are looking for periodic or recurrent phenomena, we should develop methods which can detect the presence of loops without defining explicit models. Ideally we should be able to detect the periodicity without having to develop a fully accurate model of the dynamics.

Example 2.3 Data sets can also look like this one, in which the data does not break up into disconnected groups, but instead has a structure in which there are lines (or flares) emanating from a central group.

In this case, it also suggests the presence of three distinct groups, but the connectedness of the data does not reflect this. This particular data arises from a PCR study of single nucleotide polymorphisms (SNP's).

In each of the examples above, an aspect of the *shape* of the data is what is relevant in reflecting the information about the data. Connectedness (the simplest property of shape) reflects the presence of a discrete classification of the data into disparate groups. The presence of loops in the data, another simple aspect of shape, often reflects periodic or recurrent behavior. Finally, in the third example, the shape containing flares can also suggest a classification of the data, but in such a way that the classification describes ways in which a phenomenon can deviate from the norm, which would be represented by the central core. These examples support the idea that the shape of data (suitably defined) is an important aspect of its structure, and that it is therefore important to develop methods for analyzing and understanding its shape. The part of mathematics which concerns itself with the study of shape is called *topology*, and an important theme in recent research is the adaptation of techniques from this discipline to the study of data. We will discuss this direction in this paper, beginning with a summary of topology as practiced within pure mathematics, and then discussing how the methods are adapted to the world of data.

2.2 Representing shape

Shape is a nebulous and ill-defined concept, which can be interpreted in various ways. For example, in one version of shape we might view a circle and an ellipse as being distinct, but in other forms of shape we might regard them as not being distinct since they both are loops, even though one has been stretched out of its completely round shape. Also, naive representations of shape are infinite. For example, a shape may be represented as an infinite collection of points together with a measure of distance between any pair of points.

Topology is the area of mathematics which concerns itself with studying the shape of geometric objects. The first investigation within this area was Euler's solution of the Königsberg bridge problem in 1736. It constituted the first formal recognition that there are properties of shapes which are preserved under deformations of the shape. The subject developed slowly through the 18th century and first half of the 19th, but has since then been studied intensively. One of the main goals of the field of topology is to find simple, finite, and combinatorial representations of shapes. As an example, consider the following representation of the geometry of the circle.

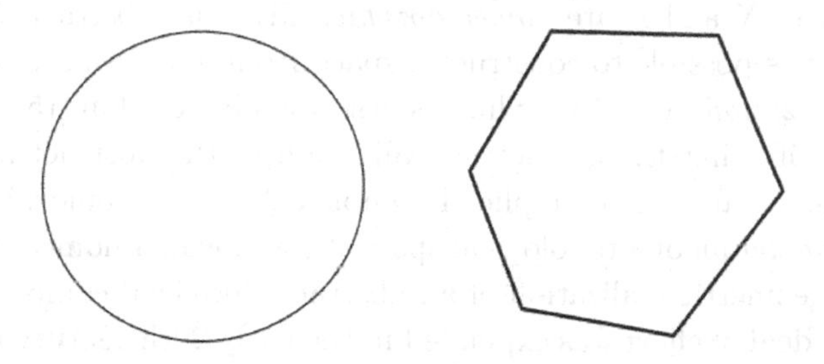

The circle is an infinite set of points. The representation of it on the right can be given as a list of six points and six edges, together with the incidence relations between them. There is some information lost in the representation, of course, namely the constant value of the curvature on the circle. An ellipse would be represented in exactly the same way as the round circle. However, the representation still retains the information that there is a loop in the space. Formally, we have the following definitions. Recall that an n-simplex in a Euclidean space $\mathbb{R}^N$ is the convex hull of a set of points $\{v_0, \ldots, v_n\}$ (the *vertices*) which are in general position. Recall that a set of $n+1$ points in $\mathbb{R}^N$ is said to be in general position if they are not contained in any affine hyperplane of dimension less than n. A *face* of a simplex is the convex hull of a subset of its vertex set. Any face of an n-simplex is a k-simplex, where $k+1$ is the cardinality of the associated set of vertices.

Definition 2.4 A *simplicial complex* is a list of simplices in some Euclidean space so that (a) any two simplices in the list intersect in a face of each of the two and (b) any face of a simplex in the list is also in the list. An *abstract simplicial complex* is a pair (V, Σ), where V is a finite set of vertices and Σ is a collection of subsets of V so that if $\sigma \in \Sigma$, and $\tau \subseteq \sigma$, then $\tau \in \Sigma$.

Remark 2.5 Given any simplicial complex X, there is a naturally associated abstract simplicial complex (V, Σ), for which V is the set of vertices of X, and so that Σ is the collection of subsets $\{v_0, \ldots, v_k\}$ so that the convex hull of $v_0, \ldots, v_n$ is one of the simplices of X. Note that an abstract simplicial complex is a purely combinatorial object, in that it is simply a list of vertices and a list of subsets of these vertices.

We recall that a continuous map $f : X \to Y$ between topological spaces is a *homeomorphism* if there is a continuous map $g : Y \to X$ so that $f \circ g = id_Y$ and $g \circ f = id_X$. If there is a homeomorphism from X

to Y, we say X and Y are *homeomorphic*. Given an abstract simplicial complex, it is possible to construct a space from it via a process of *geometric realization*, $X \to |X|$, which is described in detail in [15]. It is also easy to verify that the geometric realization of the abstract simplicial complex associated to a simplicial complex X is homeomorphic to X. By a *triangulation* of a topological space X, we mean a homeomorphism from the geometric realization of an abstract simplicial complex to X.

A great deal of effort was expended in the early 20th century in studying to what extent topological spaces are completely mirrored by associated combinatorial structures. It is clear that there are topological spaces X which are not the geometric realization of any simplicial complex. One could ask, however, if to any space which does admit a triangulation, there is an essentially unique one. This question was formulated using the notion of refinement of a simplicial complex. A simplicial complex X is a refinement of a simplicial complex Y if and only if each simplex of Y is a union of simplices of X. The complex on the left below is a subdivision of the complex on the right.

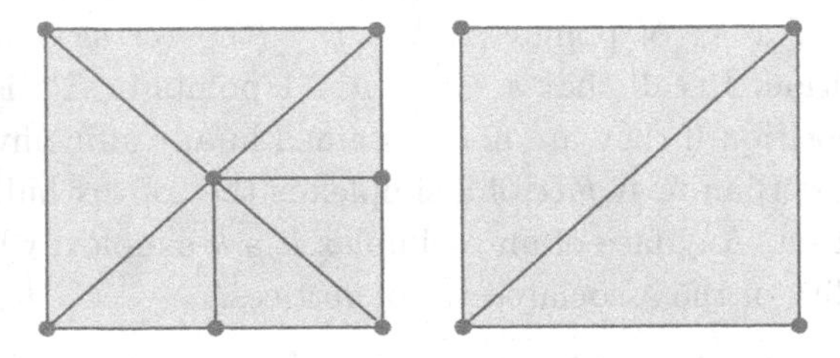

The main conjecture, called *the Hauptvermutung*, was that given two triangulations of a space X, they would admit a common refinement. The conjecture was posed in 1908, and was eventually disproved in the 1960's. The point to be made here is that questions about how topological spaces can be represented by combinatorial objects have been fundamental to the subject, and have constituted some of the greatest challenges within it.

Topologists have also developed systematic methods for representing spaces by simplicial complexes. In order to understand the statement of the main result we will be using, we first describe the notion of homotopy equivalence. Given two continuous maps $f, g : X \to Y$, a *homotopy* from f to g is a continuous map

$$H : X \times [0,1] \longrightarrow Y$$

so that $H(x, 0) = f(x)$ and $H(x, 1) = g(x)$. The intuition behind a homotopy is a one parameter family of deformations of continuous maps starting with f and ending with g, as indicated in the picture below.

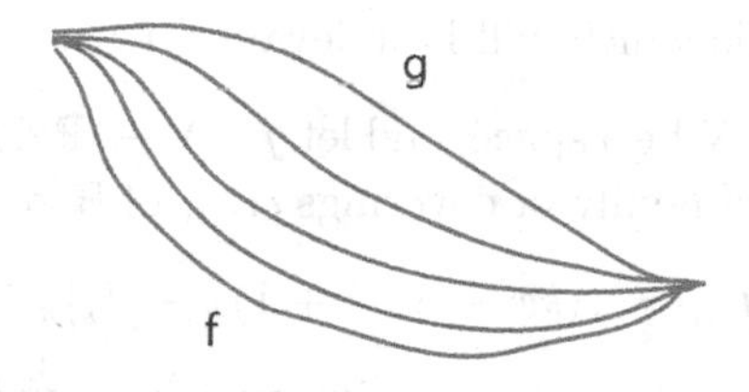

The relation of being homotopic is an equivalence relation on the set of all continuous maps from X to Y, because families of deformations can be concatenated. We may therefore speak of the *homotopy class* of a map f as the family of maps which are homotopic to f.

In this case, the space X is the unit interval $[0, 1]$, and maps from the interval into the plane are viewed as paths. The picture represents a homotopy from the lower path to the upper one. Two spaces X and Y are said to be *homotopy equivalent* if there are maps $f : X \to Y$ and $g : Y \to X$ so that the composite $f \circ g$ is homotopic to id_Y, the identity map on Y, and $g \circ f$ is homotopic to id_X. Homotopy equivalence is a much coarser notion of differentiation of shapes than homeomorphism. For example, an annulus and a circle are homotopy equivalent but not homeomorphic. In the picture below, there is a homotopy from the

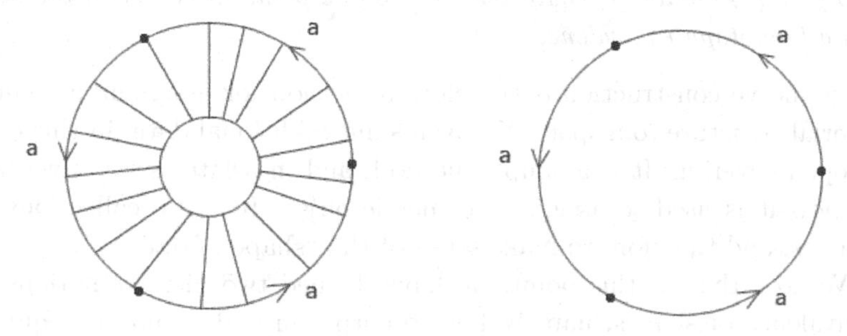

identity map on the annulus to the retraction onto the outer boundary curve which moves outward along the radial line segments.

This notion of equivalence is useful in understanding one construction of a simplicial complex from a topological space.

Definition 2.6 Let X be any set, and let $\mathcal{U} = \{U_i\}_{i \in I}$ be any covering of X, so $X = \bigcup_{i \in I} U_i$. Then by the *nerve* of $\mathcal{U}$, $N(\mathcal{U})$, we will mean the abstract simplicial complex whose vertices are the set I, and where for any non-empty $S \subseteq I$, S is a simplex in $N(\mathcal{U})$ if and only if $\bigcap_{s \in S} U_s \neq \emptyset$.

Here is an example which will be relevant for us.

Example 2.7 Let X be a space, and let $f : X \to \mathbb{R}$ denote a continuous map. We construct a family of coverings $\mathcal{U}_{x,R}$ of $\mathbb{R}$ by setting

$$\mathcal{U}_{x,R} = \{(kx - R, (k+1)x + R)\}_k.$$

Each $\mathcal{U}_{x,R}$ is a family of overlapping intervals of equal length, and with equal length overlaps. Since f is continuous, the covering

$$f^*\mathcal{U}_{x,R} = \{f^{-1}(kx - R, (k+1)x + R)\}_k$$

is also an open covering of X. Let $X_k = f^{-1}(kx - R, (k+1)x + R)$, and let $\{X_k^i\}_{1 \leq i \leq c(k)}$ denote the decomposition of X_k into its connected components, so $c(k)$ denotes the number of connected components in X_k. We now have the covering of X given by $\{\{X_k^i\}_{1 \leq i \leq c(k)}\}_k$. This covering is closely related to the notion of the *Reeb graph* of the function f. One can use multiple functions to obtain instead a map $f : X \to \mathbb{R}^n$, and cover $\mathbb{R}^n$ by overlapping n-dimensional rectangles instead.

Theorem 2.8 *Let $\mathcal{U} = \{U_i\}_{i \in I}$ be an open covering of a paracompact (see* [18] *for the definition) space X. Then there is a naturally defined homotopy class of maps $\theta : X \to |N(\mathcal{U})|$. Suppose further that for any non-empty subset $S \subset I$, we have that the subset $\bigcap_{s \in S} U_s$ is either (a) empty or (b) homotopy equivalent to the one point space. Then the map θ is a homotopy equivalence.*

The nerve construction is therefore a method for assigning a combinatorial structure to a space X, given some additional data, in this case an open covering. It is a simple method, and in a later section we will see how it is used to assign a geometric object to finite collections of samples, and therefore to make sense of the "shape of data".

We note that at this point, we have defined two distinct notions of equivalence of shapes, namely homeomorphism and homotopy equivalence. The homeomorphism classification is the more refined classification, but homotopy equivalence is often simpler to work with. We will see that loops in geometric shapes are detected by homotopy equivalence, but that flares are not.

2.3 Measuring shape

Topologists have also developed a formalism for *measuring* shape, i.e. constructing appropriate invariants which detect the presence of certain features which are not altered by continuous deformations. The idea of measurement typically involves associating real numbers to physical processes or phenomena. When dealing with shape, however, this notion of measurement is often inappropriate. The first notion of measuring shape came in the form of *Betti numbers*, which are counts of appropriately defined n-dimensional holes. That is, for every space X, there is an infinite sequence of non-negative integers $\beta_i(X)$, $i \geq 0$, called the Betti numbers of X. The integer $\beta_i(X)$ counts the i-dimensional holes in X. The intuition behind these numbers is reflected in the following values.

X	$\beta_0(X)$	$\beta_1(X)$	$\beta_2(X)$	$\beta_3(X)$
S^1	1	1	0	0
$T^2 = S^1 \times S^1$	1	2	1	0
S^2	1	0	1	0
K	1	2	1	0
M_2	1	4	1	0
S^3	1	0	0	1

The space denoted by S^1 is the usual topological circle. The usual torus can be identified with the product of two copies of S^1. The two dimensional and three dimensional spheres are denoted by S^2 and S^3 respectively. Note that the dimension is the *intrinsic* dimension of the space, so the n-dimensional sphere is the unit sphere in $n+1$-dimensional Euclidean space. The Klein bottle (denoted by K in the table) is a non-orientable surface, illustrated in the following picture (Courtesy of T. Banchoff).

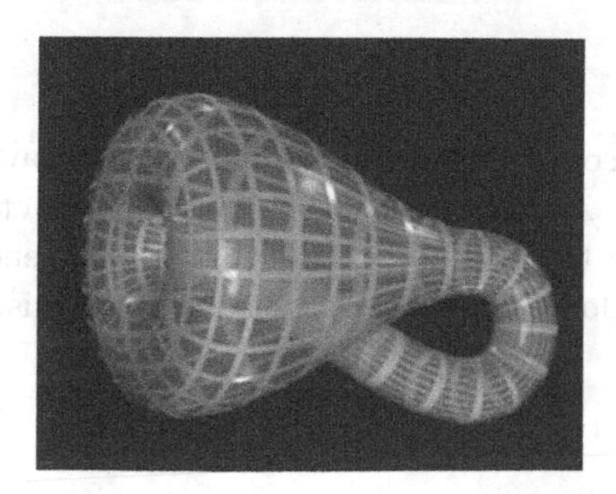

The space M_2 is the two holed torus, illustrated below. This example illustrates

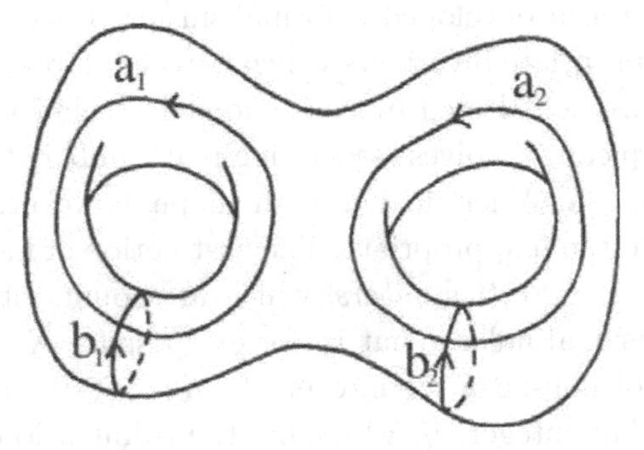

in an informal way how the Betti numbers are defined. Note that there are 4 essentially distinct loops in M_2, denoted by a_1, a_2, b_1, and b_2. This count of the number of loops is the Betti number.

We will discuss the informal ideas behind the definition of the Betti numbers. The zeroth Betti number of a space X is simply the cardinality of the number of components of X. We will therefore refer to β_0 as 0-*th order connectivity information.* We will now show that there are higher order versions of connectivity information. Consider the space X pictured below, a square in the plane with a circular hole removed.

The space X is connected, since one can clearly draw a path between any two points in X, where the path is entirely contained in X. We can ask, though, if there is a way to detect the presence of the hole or obstacle in the middle of the space. In order to do this, we consider the picture below.

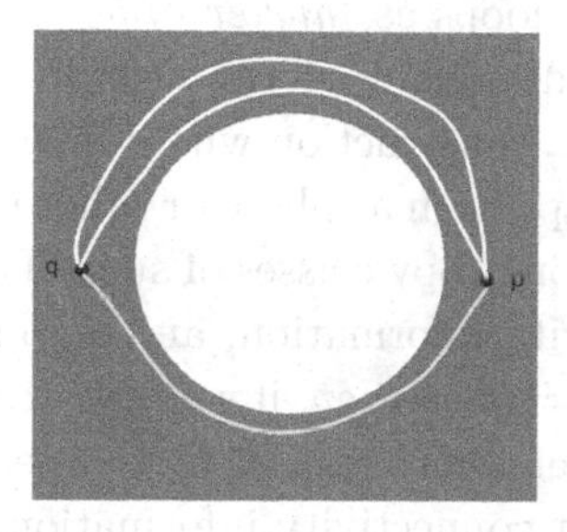

There are, of course, infinitely many paths between p and q in X. However, homotopy divides them up into many different classes. We note that the two upper paths are homotopic (via a homotopy which leaves the two points fixed), while neither of the lower two paths is homotopic to the upper path, leaving p and q fixed. Note that by concatenating any path from p to q with a path from q to p, we can obtain a *loop* in X, i.e. a map from the circle to X. There are of course many loops with initial point p, but we may group the loops by homotopy.

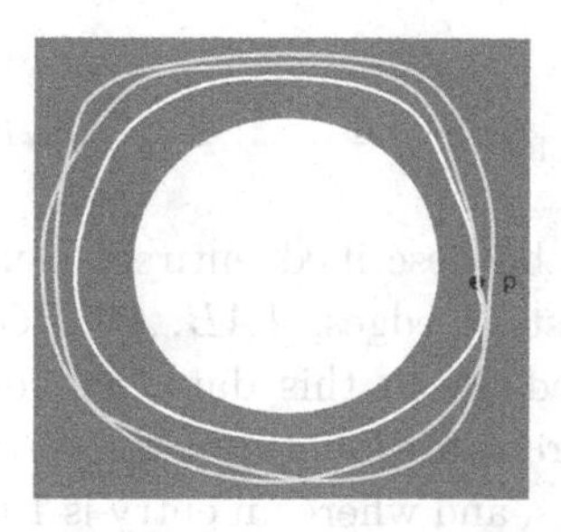

Note that the inner loop "winds around" the obstacle once, and the outer one twice. This turns out to mean that they represent different homotopy classes of loops, i.e. of maps from the circle to X. So, we are led to the idea that the classification of the loops in a space can reflect shape or structure of a space. In this case, it turns out that the homotopy classes of loops in X are in one to one correspondence with the integers, where the correspondence takes a loop to its appropriately defined "winding number". Loops which move around the obstacle in the counterclockwise direction are given negative winding numbers. We refer to this classification of loops as *higher order connectivity information*, since it reflects the redundancy of connectedness between two points. One can now imagine a construction where instead of considering loops in X, one considers maps from a sphere or other two dimensional surface into X, and consider homotopy classes of such. This would be additional higher order connectivity information, and we might refer to it as *second order connectivity information*, if we refer to connected components as zeroth order, loop classification as first order, etc. It is an obvious extension to n-th order connectivity information.

The above discussion gives an intuitive idea about how one can measure shape by using higher order connectivity information, but we have not given a satisfactory notion of how to precisely define it, or how it can be computed. These computations are turned into a linear algebra calculation in a way which is illustrated by the following example.

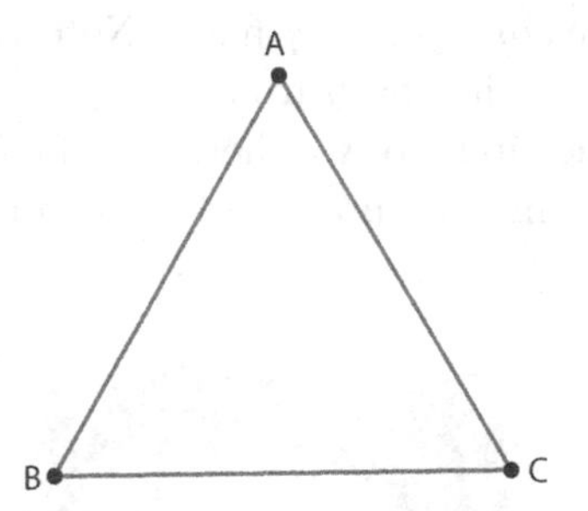

This complex can be described entirely using the list of vertices, $\{A, B, C\}$, and the list of edges, $\{AB, AC, BC\}$. Note that the full triangle is not included. From this data, we construct a matrix with Boolean entries whose rows are labeled by the vertices and whose columns are labeled by the edges, and where an entry is 1 if and only if the vertex corresponding to the row is contained in the edge corresponding to the column.

	AB	AC	BC
A	1	1	0
B	1	0	1
C	0	1	1

This matrix is referred to as the *boundary matrix* because when we regard the matrix as a linear transformation, each edge is assigned to the sum of its two endpoints, which can be thought of as the boundary if we interpret sum as union. It is an easy observation that in this case the matrix has rank 2. There are two vector spaces (over the Boolean field $\mathbb{F}_2$) we may assign to this matrix. The first is the quotient of the full 3-dimensional vector space over $\mathbb{F}_2$ by the column space of the matrix. A useful representation of this quotient would be a complementary subspace to the column space within $\mathbb{F}_2$. Such a complement would be given by the span of any single vertex. This means that the dimension of the quotient space is 1, and further examination shows that the quotient space may be identified with a Boolean vector space with basis the connected components of the complex. This suggests to us that the zeroth order connectivity information is computable from the boundary matrix, and a little bit of experimentation shows that this works for all simplicial complexes. Going back to our example, we have a second space we assign to the boundary matrix, namely the null space. It too has dimension 1, and is spanned by the element $A + B + C$. If we again permit ourselves to think of addition as union, we see that this represents the loop ABC within the simplicial complex. This suggests that just as the zeroth order connectivity information may be computed using linear algebra on a boundary matrix, first and higher order connectivity information should be computable using various kinds of linear algebra associated to the simplicial complex.

We give a brief summary of how this is carried out. Suppose we are given a simplicial complex Σ, with Σ_i denoting the set of i-simplices in Σ. For each i, we create a matrix Δ_i with rows in one to one correspondence with Σ_i, and columns in one to one correspondence with Σ_{i+1}. For each i, we also construct a Boolean vector space with basis Σ_i. The picture is now as follows.

$$\cdots \longrightarrow V_{i+1} \xrightarrow{\Delta_i} V_i \xrightarrow{\Delta_{i-1}} \cdots \xrightarrow{\Delta_1} V_1 \xrightarrow{\Delta_0} V_0.$$

Each matrix Δ_i acts as a linear transformation defined on V_{i+1} with values in V_i, from the way we defined the columns and rows. Further, it turns out that we have $\Delta_i \circ \Delta_{i+1} = 0$ for all i. This observation is the linear algebraic manifestation of the fact that the boundary of a boundary is empty. Consider a triangle, for example. Its boundary is the collection of all its faces, and when we take the boundary of each of these edges, we obtain a union (or sum) in which each vertex appears exactly twice, and is therefore zero in the Boolean field. We let N_i denote the null space of Δ_{i-1} and B_i the column space of Δ_i. Both are subspaces of

V_i, and further $N_i \supseteq B_i$. The i-th homology group $H_i(\Sigma)$ is now defined to be the quotient space N_i/B_i. The dimension of this vector space can be readily computed from the dimensions of the spaces V_i and the ranks of the matrices Δ_{i-1} and Δ_i, as

$$\dim H_i(\Sigma) = \dim V_i - \text{rank}(\Delta_i) - \text{rank}(\Delta_{i-1}).$$

This equation suggests that it is unnecessary to use the invariant terminology of vector spaces, and simply work with matrices. This is not so, because understanding the nature of homology as a vector space rather than just as a dimension is critical to (a) computing homology in many cases and (b) nearly all applications of homology. This was a key observation of Emmy Noether [8] in the mid 1920's.

So far we have discussed the computation of homology for simplicial complexes, where one is given the list of simplices making up the complex. Of course, what is really desirable is a definition which can be applied to any topological space. Even if the space is homeomorphic to a simplicial complex, one would like to have a definition which doesn't depend on a particular structure. It turns out that this is possible, due to the work of Eilenberg [11], who defined so-called *singular homology.* In rough terms, to define the homology groups of a space X, Eilenberg defined an object with infinitely many k-simplices, one for each continuous map of the standard k-simplex into X. He then managed to prove various properties of the construction, including the fact that it produces results which are isomorphic to the simplicial calculations described above whenever the space is equipped with the structure of a simplicial complex. Of course, such infinite constructions cannot be computed directly by a machine, but the homology groups enjoy certain properties which make indirect computations very feasible. One of these properties is functoriality, which we have already discussed, and another is excision, which allows one to understand the behavior of homology groups under unions. The summary of the outcome of Eilenberg's work is the following.

- For every topological space X and non-negative integer k, there is a Boolean vector space $H_k(X)$.
- The construction $H_k(-)$ is *functorial*, i.e. for every continuous map $f : X \to Y$ there is an associated linear transformation

$$H_k(f) : H_k(X) \to H_k(Y)$$

so that if we have continuous maps $f : X \to Y$ and $g : Y \to Z$, then $H_k(g \circ f) = H_k(g) \circ H_k(F)$.

- The construction of $H_k(f)$ is *homotopy invariant* in the sense that if we are given two homotopic maps $f, g : X \to Y$, then $H_k(f) = H_k(g)$ as linear transformations from $H_k(X)$ to $H_k(Y)$.

Remark 2.9 It follows from the homotopy invariance that homotopy equivalent spaces have isomorphic homology groups, and therefore that homology is capable only of distinguishing spaces which are not homotopy equivalent, and will not distinguish between non-homeomorphic spaces which are homotopy equivalent. This means that it can distinguish a circle from an interval, but not a letter "Y" from a letter "J".

2.4 Topology and the shape of data

In Section 2.2, we have introduced certain mechanisms for representing the shape of actual geometric objects via simplicial complexes. We want to represent the shape of data sets in a similar way. One mechanism for representing coarse aspects of shape is *cluster analysis* [14]. Cluster analysis consists of a family of different methods for partitioning data sets into conceptually coherent pieces. Typically, they take as their starting point a distance function d on the data set X. There are a number of interesting methods.

1. **k-means clustering:** In this method, one assumes one is given a number k, and attempts to divide the data set into k clusters. The data set is assumed to be Euclidean, so we may compute the centroid of any set of data points. The assigned partition is defined to be the solution to an optimization problem, for which the objective function is

$$\sum_{j=1}^{k} \sum_{i=1}^{n} |x_i^{(j)} - c_j|^2$$

where the c_j's are the centroids of the k-different clusters. The method proceeds by first initializing the centroids using some heuristic scheme, then clustering by assigning each point to its nearest centroid. One then computes the centroids of each cluster so defined, and then reclusters using the computed centroids. This method proceeds until one obtains stability.

2. **Single linkage clustering:** In this method, one takes as a given a scale parameter ϵ, and forms the graph in which the vertices are the elements of X, and where two points are connected by an edge if and only if their distance is less than or equal to the value of ϵ. One then forms the connected components of the graph, which gives a partition of the graph (and therefore the vertices).
3. **Hierarchical single linkage clustering:** This method builds on single linkage by recognizing that whenever $\epsilon \leq \epsilon'$, the partition associated to ϵ is finer than the partition associated to ϵ', and that one therefore obtains a mapping from the set of clusters at the scale ϵ to those of ϵ'. This collection of maps of clusterings gives rise to a tree structure on X, together with a map from the tree to the real line. This is called a *dendrogram*. The dendrogram, rather than being a single choice of partitioning, gives a useful representation of the clusterings at all values of the scale parameter ϵ.
4. **General linkage methods:** One supposes that one is given a rule which assigns a number to any pair of subsets $S, T \subseteq X$. It might be

$$\min_{s \in S, t \in T} d(s,t),$$

or the average distance between points in S and T, or it might be

$$\max_{s \in S, t \in T} d(s,t).$$

If we denote the linkage function by λ, the algorithm proceeds by beginning with the discrete partition, in which each point is its own block in the partition. It then finds the minimal non-zero value of $\lambda(x,y)$, for $x, y \in X$, and merges any pair of points which achieve this minimal value. This produces a new partition of X, as well as a value ϵ_1, namely the minimal value of λ which was used in merging the points. Next, we find the minimal value of $\lambda(C_1, C_2)$ over the clusters computed at the previous step, and merge all clusters achieving the minimal value, as well as a threshold value ϵ_2, the minimal value of λ which was used. This step can now be repeated, and one obtains a sequence of partitions $\mathfrak{P}_i$ of X, together with associated values ϵ_i of λ. We also note that $\mathfrak{P}_i$ is finer than $\mathfrak{P}_{i+1}$, and that we therefore obtain a dendrogram just as we did in the single linkage case. The schemes constructed using the linkage functions defined above are called hierarchical single linkage, average linkage, and complete linkage, respectively.

We observe that, in a sense, clustering methods on a data set X are analogous to the connected components $\pi_0(X)$ for a space X. What is interesting is that although in topology, there are two notions of connected components, namely connected components and path connected components, in the study of data there are actually many, and further that one often assigns a whole array of partitions of increasing coarseness to a data set.

Another method for attempting to represent the shape of data is to somehow find a projection of the data set into a Euclidean space of two or possibly three dimensions. One can then view the projection of the data set in the relevant Euclidean space, and possibly draw conclusions concerning its shape. We will refer to such methods as *scatterplot methods*. There are several such methods.

1. **Projection pursuit:** This method assumes the data is itself Euclidean, i.e. consists of vectors of a fixed length. Each coordinate then provides a function on the data, and one can consider the distribution of values of that coordinate. The presumption is that one has a quantitative measure of how "informative" each coordinate is. For example, kurtosis and/or skewness of the distribution might be a useful measure, or measures of the degree to which the distribution is multimodal. One then selects the two or three coordinates which maximize this measure, and projects to the Euclidean space using these coordinates. Visual inspection of this projection can often be used to identify aspects of the shape of the data.
2. **Principal component analysis:** This method again assumes Euclidean data, and proceeds as follows. The data set can be represented as a data matrix, with the rows corresponding to the data points and the columns to the coordinates or fields defining the data. The *singular value decomposition* asserts that for every $m \times n$ matrix M, there are orthogonal matrices U and V so that UMV is of the form

$$\left[\begin{array}{c|c} D & 0 \\ \hline 0 & 0 \end{array}\right]$$

where D is a diagonal matrix of the form

$$\begin{bmatrix} d_1 & 0 & \cdots & 0 \\ 0 & d_2 & \cdots & 0 \\ \vdots & \vdots & \ddots & \vdots \\ 0 & 0 & \cdots & d_k \end{bmatrix}$$

where $d_1 \geq d_2 \geq d_3 \geq \cdots \geq d_k$. The d_i's are referred to as the *singular values*, and are uniquely determined by M. The matrix V provides a coordinate change to the data set in which the first few coordinates frequently carry the most information about the data set. The first two or three principal components then provide a two or three dimensional scatterplot for the data, which again can often be used to identify aspects of the shape of the data visually.

3. **Multidimensional scaling:** This method can be applied to non-Euclidean data, in the form of a distance matrix which records the distances between pairs of points in a symmetric matrix. The method proceeds by selecting a fixed n, and determining via an optimization procedure a map to n-dimensional Euclidean space which minimizes a notion of distortion of the metric, comparing the metric coming from the data matrix with the metric obtained from the embedding of the data set in Euclidean space. The method produces an ordering of the coordinates in such a way that one often obtains small distortion among a small number of initial coordinates. Again, the method produces a scatterplot from which features can often be found.

One disadvantage of the scatterplot methods is that they produce a Euclidean representation in which every data point occurs. For large data sets, this means that the scatterplot is often complicated and features are difficult to discern. In addition, it becomes difficult to apply automatic methods for feature location. We will demonstrate a method which combines the advantages of clustering methods with those of the scatterplot methods to obtain a compressed representation of the data in terms of simplicial complex. The basis for our construction will be a direct adaptation of the method described in Example 2.7.

We suppose we are given a data set X equipped with a metric. Suppose that we have functions

$$\{f_i : X \to \mathbb{R}\}_{1 \leq i \leq k}.$$

We also fix coverings $\mathcal{U}_i$ of $\mathbb{R}$, each of the form $\mathcal{U}_{x_i,R_i}$, as in Example 2.7. We write $\mathcal{U}_i = \{U_{\alpha_i}\}_{\alpha_i \in A_i}$. Then for every k-tuple $\vec{\alpha} = (\alpha_1, \alpha_2, \ldots, \alpha_k)$, with $\alpha_i \in A_i$, we obtain a subset $X_{\vec{\alpha}} \subseteq X$. The subsets $X_{\vec{\alpha}}$ clearly cover X. At this point, the process has been entirely in parallel with the construction in Example 2.7, in that we have covered X with overlapping "buckets". This "bucketing" procedure depends only on the functions f_i and the coverings $\mathcal{U}_i$, and not on the fact that X carries a topology. However, the next step in the process described in Example 2.7 involves the taking of connected components of each of the buckets, and this does not have an obvious extension here since the spaces involved have the discrete topology. This means that we must find a suitable replacement, which partitions each set $X_{\vec{\alpha}}$ into disjoint pieces. Here we substitute a clustering algorithm. We don't specify which one, since any one can be used. One could, for example, choose single linkage clustering with a fixed parameter, or versions of k-means, or average or complete linkage with a fixed parameter as well. One could even permit the use of a linkage method with parameter varying with the bucket, using some heuristic scheme. This is what is adopted in the version of the methodology used in [21], [2], [20], and other applications of the method. Once we have formed a clustering

$$\coprod_{j=1}^{n_{\vec{\alpha}}} X_{\vec{\alpha}}(j)$$

of $X_{\vec{\alpha}}$, we form the nerve of that covering of X, and this is our geometric representation of the data set X. We will now demonstrate that the method is useful in calling attention to meaningful geometric features in data sets, by providing numerous examples.

Example 2.10 This example came up in the analysis of simulated RNA hairpin folding. The simulations were constructed by the Folding@home project led by Vijay Pande at Stanford. In this case, the data set consisted of conformations of a fixed molecule, and as reference map we used a suitably chosen proxy for density.

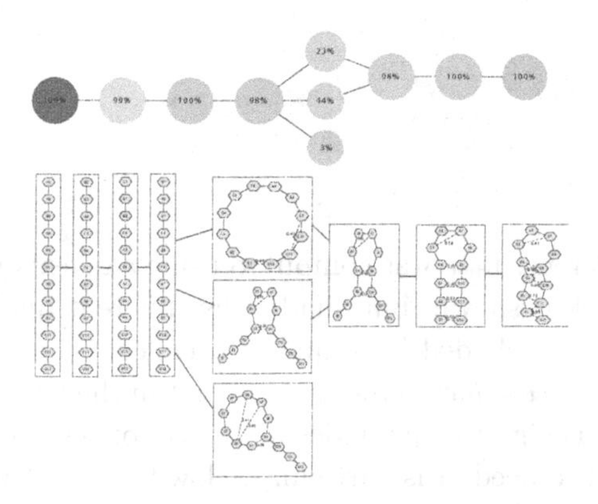

The output of the algorithm was the upper network. One important feature is the small loop in the center. Ones attention is called to it, but it may of course be an artifact. However, examination of the two different nodes on opposite sides of the loop turned out to contain conformations which represented distinct trajectories to the folded state. The conformations are themselves illustrated in the diagram below the upper network. We note that the loop is actually a rather small feature in a larger data set. In this case, clustering methods by themselves would not be able to find this feature, both because the space appears to be connected, but also because of the scale of the feature. On the other hand, small features such as this one are often "washed out" by scatterplot methods, and this is indeed what happens in this case. See [2] and [24] for a thorough discussion of examples of this type.

Example 2.11 This example works with the data from the Miller-Reaven diabetes study [17], consisting of 145 patients, with coordinates coming from various metabolic measurements and a normalized notion of weight.

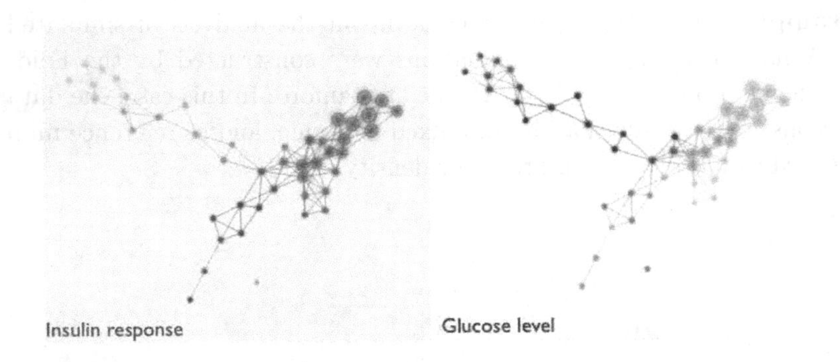

Again, the reference map was computed using a density estimator, and the same network is shown, but shaded by different quantities in each case. The left one is shaded by a quantity called *insulin response*, and the right one by *glucose level*. One can see that in the leftmost image, the insulin response is high for the nodes at the tip of the bottom flare, and lower for the rest. Indeed, it is particularly low for the leftmost flare. The right hand image shows the same network, but with each node colored by the average value of the glucose level. We now see that the high values of glucose level occur at the tip of the leftmost flare, and that the lower values occur in the bottom flare. The nodes in the upper flare consist mostly of healthy patients, the lower flare of "pre-diabetics", and the left hand flare of "overt diabetics". Pre-diabetics are characterized by strong insulin response and normal glucose levels, and overt diabetics by very weak insulin response and high glucose levels.

Example 2.12 This example was constructed using a microarray study of gene expression in breast cancer, and is published in [23]. It contains 262 tumors, and the coordinates are the expression levels for a collection of genes of highest variation among the ones provided by the study. In an analysis of this data set (see [20]), a power of distance from the origin in the *Disease Specific Genomic Analysis* (DSGA) transform of the data was used as the reference map. This transform, introduced in [19], has the effect of coordinatizing the data set in such a way that the variation of profiles of healthy patients are deemphasized, providing a coordinatization which is more reflective of deviation from normal expression profiles.

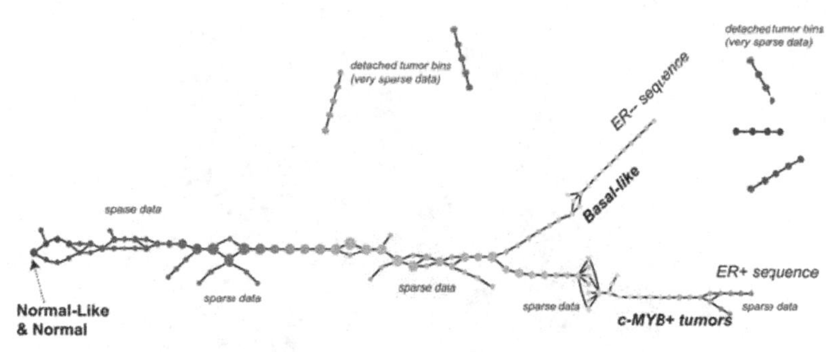

The network is colored by the value of the reference map. Hence, the nodes on the left include some normal tissue samples, and the group of tumors approximating that normal behavior form a "normal-like" group, which has very good survival. On the right, as one moves further away from the normal behavior, one sees two flares. The upper one, labeled "basal-like", consists of tumors which have low estrogen receptor expression. There are particular genes correlated with estrogen receptors, and low values of their expression is known to be correlated with bad prognosis. The lower flare consists of another group of tumors, whose expression profile is far from normal. However, it turns out that if one considers the members of that flare beyond the area labeled "sparse data", one finds that all members of that group survived the length of the study. These patients have high estrogen receptor expression levels, but are not exclusively characterized by it. Other genes involved in defining this group are reported in [20].

Example 2.13 This is an example which shows that even much less well-defined features of a representation of the data can correspond to interesting information concerning the data. The data set in question is the so-called Hapmap study (see [12]). In it, sequence information was obtained from three separate populations, one consisting of Han Chinese and Japanese, one consisting of members of the Yoruba group in Nigeria, and the third people of European ancestry, residing in Utah. The analysis below uses measures of copy number variation from this data set. Copy number variations are alterations of the DNA of a genome that results in the cell having an abnormal number of copies of one or more sections of the DNA, and copy number variation also refers to a number which measures the degree to which a certain segment of the genome is abnormal in this way. The image below is what is obtained using a measure of centrality as reference maps.

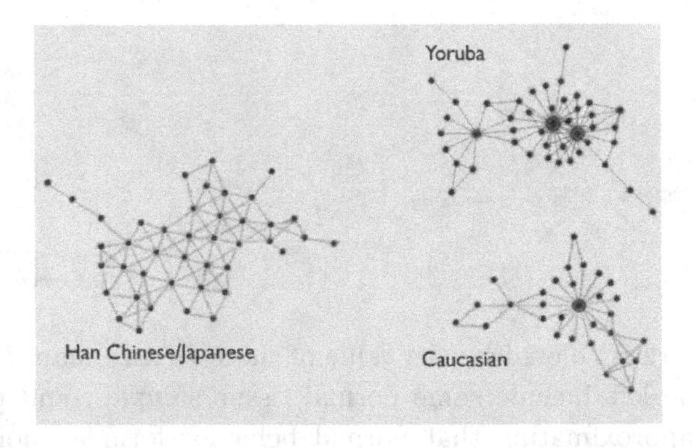

In this case, it is evident that the data breaks up into three distinct pieces, corresponding to the three different groups. This is an easy result to obtain, and can be duplicated by standard clustering methodology or a scatterplot method. What is interesting, though, is the presence in the Yoruba and European groups of starlike groups. These turn out to be due to the presence of parent-child pairs, something which was permitted in the Yoruba and European studies but not in Oriental study. The point here is that the method we have constructed can be used in all phases of data analysis, even the early phases which might be considered closer to quality control than analysis.

2.5 Measuring the shape of data

Finite data sets do not carry any interesting topology, since they are always discrete as topological spaces. This means that we must associate topological spaces to them. The key ingredient will always be a distance function on X. We will take our guidance from the single linkage clustering methods discussed in Section 2.2. Recall that this method begins with a finite metric space X and a threshold ϵ, constructs the graph whose set of vertices is X, and where two points x and y are connected by an edge if and only if $d(x, y) \leq \epsilon$. It then proceeds to construct the connected components of the graph, and we obtain the partition of the vertex set from this decomposition. The important observation is that this graph can actually be viewed as a simplicial complex, in which a $(k+1)$-tuple $\{x_0, x_1, \ldots, x_k\}$ spans a simplex if and only if all the pairwise distances $d(x_i, x_j)$ are $\leq \epsilon$. This simplicial complex is called the *Vietoris-Rips complex* with scale parameter ϵ, and is denoted by $VR(X, \epsilon)$. A typical such complex is shown below. It is constructed from nine points sampled from an annulus. The value of the scale parameter ϵ is indicated by the length of the segment below the annulus. Note that all the segments drawn within the annulus are shorter than the reference segment below, which is why they are included. Note that non-adjacent points are further apart than ϵ.

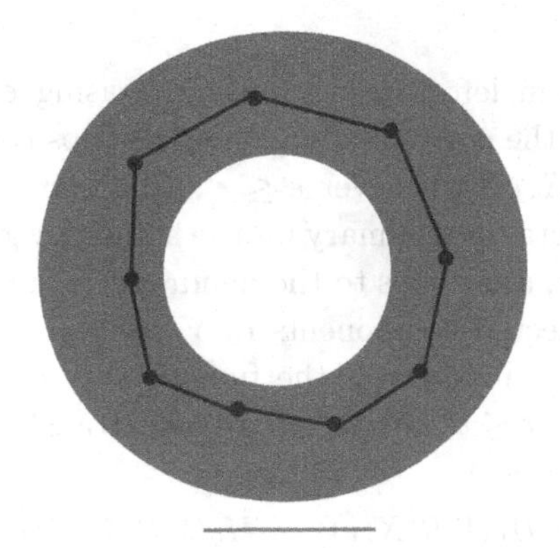

We see that in this example, the Vietoris-Rips complex is homotopy equivalent to the space from which it was sampled, the annulus. It follows that the homology of the Vietoris-Rips complex will be identical to that for the annulus. This will not always happen, of course. For example, if ϵ is less than the smallest interpoint distance occurring between pairs of distinct sampled points, there will be no edges and the space will be homotopy equivalent to nine discrete points. On the other hand, if the distance is larger than the diameter of the annulus, then all possible simplices will be included, and the resulting simplex will have the homotopy type of a point, not that of a circle or annulus. These observations suggest that one needs to choose the value of ϵ cleverly in order to obtain the topological type of the space correctly. This problem is very analogous to the problem one faces in choosing a scale parameter value for single linkage clustering, and it is not in general solvable. This suggests that one attempt to adopt a philosophy similar to the hierarchical clustering procedure adopted in many statistical investigations. The variant of the hierarchical method which applies to homology is called *persistent homology*, and we will discuss briefly how it works.

The key observation is that as one lets ϵ increase, one obtains a nested family of simplicial complexes, as in the picture below.

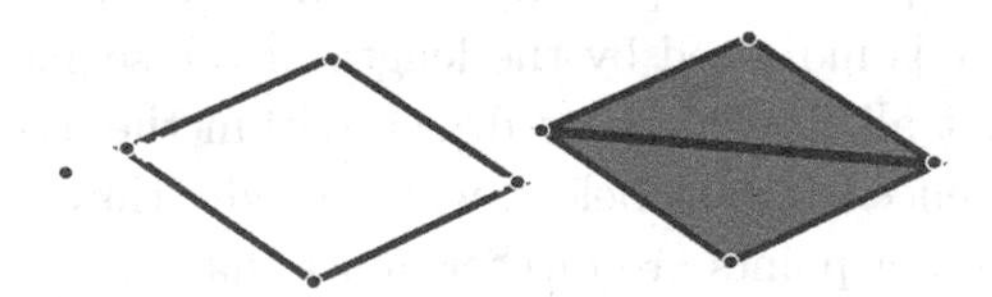

As we move from left to right with increasing ϵ, we note that we have inclusions of the corresponding Vietoris-Rips complexes. Formally, $VR(X, \epsilon) \subseteq VR(X, \epsilon')$ whenever $\epsilon \leq \epsilon'$. The question is now whether there is a way to obtain a summary for the homology groups $H_k(VR(X, \epsilon))$ over all values of ϵ, analogous to the dendrogram summary for the clusters, i.e. the connected components of $VR(X, \epsilon)$. The key to the construction of such a summary is the functoriality of the homology construction. For each $\epsilon \leq \epsilon'$, we obtain a linear transformation of Boolean vector spaces

$$H_k(VR(X, \epsilon)) \to H_k(VR(X, \epsilon')).$$

What this tells us is that we now obtain, for each finite metric space X and each $k \geq 0$, an object which consists of a family of Boolean vector spaces $\{V_\epsilon\}_\epsilon$ together with linear transformations

$$V_\epsilon \to V_{\epsilon'}$$

whenever $\epsilon \leq \epsilon'$. This kind of algebraic object is called a *persistence vector space*, and it turns out they admit an explicit classification, just as finite dimensional vector spaces over a field are classified by their dimension. For any pair of real numbers (r, s), we define the *cyclic persistence vector space associated to* (r, s) to be the persistence vector space $\{V_\epsilon\}_\epsilon$, where $V_\epsilon = \{0\}$ for $\epsilon < r$ or $\epsilon \geq s$, $V_\epsilon = k$ for $r \leq \epsilon < s$, and where all the linear transformations $V_\epsilon \to V_{\epsilon'}$ are identity transformation whenever $r \leq \epsilon < s$.

Theorem 2.14 *The persistence vector spaces associated to the increasing sequences of Vietoris-Rips complexes for a finite metric space are isomorphic to a finite direct sum of cyclic persistence modules for various values of* (r, s). *Further, the summands occurring are uniquely defined, up to reordering.*

This result is proved in [25]. It relies on the classification of modules over a principal ideal domain (see [9], [25]). Each of the parameter pairs (r, s) defines a unique closed interval on the real line, and we can therefore parametrize the isomorphism classes of persistence vector spaces (with certain finiteness hypotheses, satisfied for the persistence modules associated to finite metric spaces) by finite families of intervals on the real line. We call such a family a *bar code*. Each finite metric space has such a bar code for every integer $k \geq 0$. Here is a typical bar code assigned to a finite metric space pictured below. This is a one-dimensional bar code, representing the presence of loops in the data set.

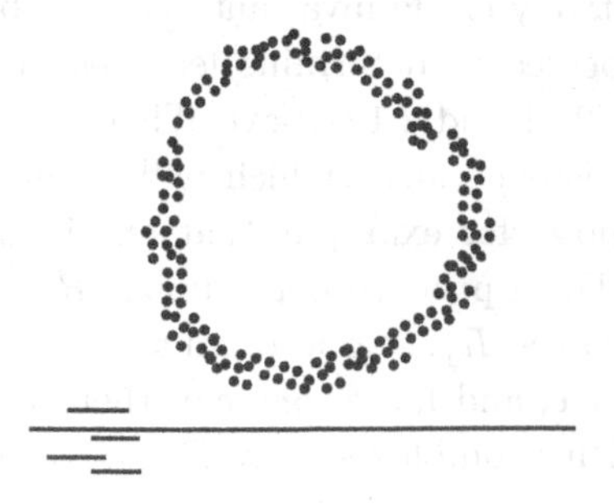

Note that there is a single long bar, and a number of shorter ones. In the case of data sets, one does not obtain a single integer, but rather this kind of profile of "features" (loops in this case) of varying lengths, reflecting the extent to which the feature survives across changes of scale. The rough interpretation is that long bars correspond to actual geometric features which might be present in a space from which the set was sampled, and the short ones correspond to noise or small failures of the set to approximate the space. These barcodes can be interpreted as a "signal" for the presence of certain kinds of geometric behavior within the data set. They can also reflect more subtle kinds of geometric behavior, such as hierarchy, which is pictured below.

The stepwise structure in the bar code reflects the hierarchical structure in the data set.

These methods have been applied to various different kinds of data. In [4] and [1], persistent homological techniques were used to identify the structure of spaces of small square patches in natural images, in [22] to study the behavior of arrays of neurons in the primary visual cortex of Macaque monkeys, and in [16] to study the structures of branched polymers.

Homology is a relatively crude invariant, as was observed in Remark 2.9. It cannot be expected to determine features in data sets such as the ones in Examples 2.11 and 2.12 above. There are modifications one can make to the persistence ideas which will permit the detection of such structures. Suppose, for example, that one is given a real valued function $f : X \to \mathbb{R}$. For a positive real number R, we let $X[R]$ denote the subset $\{x \in X | f(x) \leq R\}$. If one permits oneself to select a value of the scale parameter ϵ, and fix it, one can then obtain an increasing sequence of Vietoris-Rips complexes $V(X[R], \epsilon)$, so that

$$VR(X[R], \epsilon) \subseteq VR(X[R'], \epsilon)$$

whenever $R \leq R'$. Applying H_k to this construction yields a persistence vector space, from which one can compute bar codes. A particularly interesting function on X is

$$\mathfrak{C}(x) = \frac{1}{\sum_{x' \in X} d(x, x')}.$$

It is a measure of *centrality*, so that large values of this function correspond to points which are more central in the data set, and smaller values correspond to more peripheral points. Given an data set such as the one below,

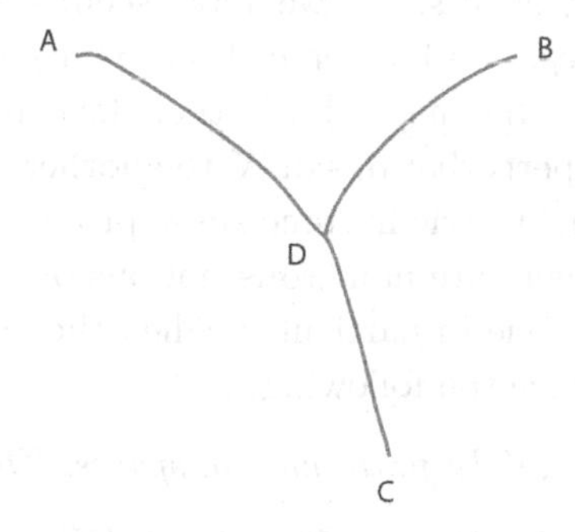

$\mathfrak{C}$ will have small values at the peripheral points A, B, and C, and larger values as one approaches the center D. This means that for small values c of $\mathfrak{C}$, we will have three distinct components in $VR(X[c], \epsilon)$, and that they will merge for larger values of $\mathfrak{C}$. This will mean that we have a zero-dimensional barcode of the form

The longer bar is actually understood to be infinite. We see that there are three components for an initial segment, and that two of the bars vanish, reflecting the merge of the three components. In this way, one can modify the way one does persistence so as to permit discrimination of shapes which are homotopy equivalent but not homeomorphic.

Remark 2.15 Of course, the fact that we are making a choice of the scale parameter ϵ is awkward, and it is not clear how to make this choice. What would be extremely useful is a method for studying the profile of components of the Vietoris-Rips complexes as *both* ϵ and $\mathfrak{C}$ vary. There is work in this direction, under the heading of *multidimensional persistence*. It is described in [5] and [6].

Persistence barcodes are a fundamental tool for studying finite metric spaces. One basic result is the stability theorem for barcodes, proved by F. Chazal, D. Cohen-Steiner, L. Guibas, F. Mémoli and S. Oudot in [7]. To state this result, we need to recall some constructions. In [13], M. Gromov constructed a metric on the family of all compact metric spaces, called the *Gromov-Hausdorff metric*, denoted d_{GH}. It of course applies to finite metric spaces. As we have seen, for any finite metric space X and non-negative integer k, there is a persistence barcode $\mathfrak{B}_k(X)$ associated to X. One may ask whether the construction $\mathfrak{B}_k$ is in any sense stable, i.e. if perturbations of X to another metric space close in the Gromov-Hausdorff metric in some sense produce small changes in $\mathfrak{B}_k$. It turns out that there are numerous notions of distance on the set of persistence barcodes, one in particular called the *bottleneck distance*, denoted d_B. We then have the following.

Theorem 2.16 *Let X, Y be finite metric spaces. Then we have*

$$d_{GH}(X,Y) \geq d_B(\mathfrak{B}_k(X), \mathfrak{B}_k(Y)).$$

The Gromov-Hausdorff distance is notoriously difficult to compute, so this theorem provides a readily computable lower bound for d_{GH}, an important theoretical observation. The theorem is also important from the point of view of applications, where it is understood that all data sets are subject to noise, and that it is therefore very important to know that small perturbations of the input data will give small perturbations of invariants one constructs.

Acknowledgement. This work was supported by AFOSR Grant FA9550-09-1-0643, Princeton Subaward #00001716-2.

References

[1] H. Adams and G. Carlsson, On the non-linear statistics of range image patches. *SIAM Journal on Imaging Sciences* **2**, 110–117, 2009.

[2] G. R. Bowman, X. Huang, Y. Yao, J. Sun, G. Carlsson, L. J. Guibas and V.S. Pande, Structural insight into RNA hairpin folding intermediates. *J. Amer. Chemical Soc. Comm.*, **130**, 9676–9678, 2008.

[3] G. Carlsson, Topology and data. *Bull. Amer. Math. Soc.*, **46**, 255–308, 2009.

[4] G. Carlsson, T. Ishkanov, V. de Silva and A. Zomorodian, On the local behavior of spaces of natural images. *International Journal of Computer Vision*, **76**, 1–12, 2008.

[5] G. Carlsson and A. Zomorodian, The theory of multidimensional persistence. *Discrete & Computational Geometry*, **42**, 71–93, 2009.

[6] G. Carlsson, G. Singh and A. Zomorodian, Computing multidimensional persistence. in *Algorithms and Computation*, Lecture Notes in Computer Science, **5878**, 730–739, 2009.

[7] F. Chazal, D. Cohen-Steiner, L. Guibas, F. Mémoli and S. Oudot, Gromov-Hausdorff stable signatures for shapes using persistence. *Computer Graphics Forum*, **28**, 1393–1403, (2009).

[8] A. Dick, *Emmy Noether: 1882-1935.* Boston: Birkhäuser, (1981), Translated by H. I. Blocher.

[9] D. Dummit and R. Foote, *Abstract Algebra.* Third edition. John Wiley & Sons, Inc., Hoboken, NJ, 2004.

[10] H. Edelsbrunner, D. Letscher, and A. Zomorodian, Topological persistence and simplification. *Discrete and Computational Geometry*, **28**, 511–533, 2002.

[11] S. Eilenberg, Singular homology theory. *Ann. of Math.* **45**, 407–447, 1944.

[12] R. Gibbs et al, The international HapMap project. *Nature*, **426**, 789–796, 2003.

[13] M. Gromov, *Metric Structures for Riemannian and Non-Riemannian Spaces.* Progress in Mathematics, vol. 152. Birkhäuser Boston Inc., Boston, MA, 1999.

[14] J. Hartigan, *Clustering Algorithms.* Wiley Series in Probability and Mathematical Statistics. John Wiley & Sons, New York-London-Sydney, 1975.

[15] A. Hatcher, *Algebraic Topology.* Cambridge University Press, Cambridge, 2002.

[16] R. MacPherson and B. Schweinhart, Measuring shape with topology. arXiv:1011.2258v2, 2010.

[17] R. Miller, Discussion - projection pursuit. *Ann. Statistics* **13**, 510–513, 1985.

[18] J. Munkres, *Topology: A First Course.* Prentice-Hall, Inc., Englewood Cliffs, N.J., 1975.

[19] M. Nicolau, R. Tibshirani, A.-L. Børresen-Dale and S. Jeffrey Disease-specific genomic analysis: Identifying the signature of pathologic biology. *Bioinformatics*, **23**, 957–965, 2007.

[20] M. Nicolau, A. Levine and G. Carlsson, Topology based data analysis identifies a subgroup of breast cancers with a unique mutational profile and excellent survival. *Proc. National Acad. Sciences*, **108**, 7265–7270, 2011.

[21] G. Singh, F. Memoli and G. Carlsson, Topological methods for the analysis of high dimensional data sets and 3D object recognition. *Point Based Graphics 2007*, Prague, September 2007.

[22] G. Singh, F. Memoli, T. Ishkhanov, G. Sapiro, G. Carlsson and D. Ringach, Topological structure of population activity in primary visual cortex. *Journal of Vision*, **8**, 1–18, 2008.

[23] M. van de Vijver et al. A gene-expression signature as a predictor of survival in breast cancer. *New Engl. J. Med.*, **347**, 1999–2009, 2002.

[24] Y. Yao, J. Sun, S. Huang, G. Bowman, G. Singh, M. Lesnick, L. Guibas, V. Pande and G. Carlsson, Topological methods for exploring low-density states in biomolecular folding pathways. *Journal of Chemical Physics*, **130**, 144115, 2009.

[25] A. Zomorodian and G. Carlsson, Computing persistent homology. *Discrete and Computational Geometry*, **33**, 249–274, 2005.

3

Upwinding in Finite Element Systems of Differential Forms

Snorre Harald Christiansen

Centre of Mathematics for Applications & Department of Mathematics
University of Oslo

Abstract

We provide a notion of finite element system, that enables the construction of spaces of differential forms, which can be used for the numerical solution of variationally posed partial differential equations. Within this framework, we introduce a form of upwinding, with the aim of stabilizing methods for the purposes of computational fluid dynamics, in the vanishing viscosity regime.

Foreword

I am deeply honored to have received the first Stephen Smale prize from the Society for the Foundations of Computational Mathematics.

I want to thank the jury for deciding, in what I understand was a difficult weighing process, to tip the balance in my favor. The tiny margins that similarly enable the Gömböc to find its way to equilibrium, give me equal pleasure to contemplate. It's a beautiful prize trophy.

It is a great joy to receive a prize that celebrates the unity of mathematics. I hope it will draw attention to the satisfaction there can be, in combining theoretical musings with potent applications. Differential geometry, which infuses most of my work, is a good example of a subject that defies perceived boundaries, equally appealing to craftsmen of various trades.

As I was entering the subject, rumors that Smale could turn spheres inside out without pinching, were among the legends that gave it a sense of surprise and mystery. I also remember reading about Turing machines built on rings other than $\mathbb{Z}/2\mathbb{Z}$, which, together with parallelism and quantum computing, convinced me that the foundations of our subject were still in the making. Happy for the occasion provided by the FoCM conference, to meet the master, I was also a bit intimidated to learn that we have a common interest in discrete de Rham sequences. They are the topic of this paper.

Many people have generously shared their insights and outlooks with me. I feel particularly indebted, mathematically as well as personally, to Jean-Claude Nédélec, Annalisa Buffa and Ragnar Winther. I'd like to dedicate this paper to my grandmother, who will probably joke, as she usually does, that she has noticed some mistakes on page 5. She constitutes an important fraction of my readership, also numberwise.

I'm very grateful for this opportunity to make my work more widely known.

3.1 Introduction

Finite elements come in a variety of brands. The particular flavor considered here, called mixed finite elements, was pioneered by [36, 34] and has since developed into a versatile tool for the numerical solution of a variety of partial differential equations describing, for instance, fluid flow and electromagnetic wave propagation [9, 37, 31]. As remarked in [7], lowest order elements correspond to constructs in algebraic topology referred to as Whitney forms [44, 45, 23]. Uniting these strands has led to the topic of finite element exterior calculus [27, 2, 28, 15, 3], most recently reviewed in [5].

Finite element systems (FES) were introduced in [16] to provide a generalization to arbitrary dimension of the dual elements constructed in [14]. This general framework allows for a unified analysis of the preceding mixed finite elements, but can also accommodate polyhedral decompositions of spaces (rather than just simplices and products thereof) as well as general differential forms (rather than just polynomial ones). This paper contains an introduction to FES, referring to [17, 20] for more ample treatments. The flexibility of FES with respect to meshing techniques is already quite standard in mimetic finite differences and finite volumes, with which one observes a confluence of techniques [11, 6, 24]. For our present purposes it is, however, the flexibility with respect to choice of local functions that is of interest, since we will need exponentials and variants thereof.

In spite of the topological twist of the subject, which might remind us of [25, 8], the "upwinding" of the title refers not to winding numbers but to wind, of the blowing kind. The equations of fluid dynamics typically feature a convective term, hyperbolic in nature, moderated by a diffusive one, elliptic in nature. As the viscosity vanishes, the nature of the equations changes, say from Navier-Stokes to Euler's equations, in what is called a singular limit. The highest order derivatives are eliminated in a delicate limiting process, where fields display sharp gradients and form boundary layers.

To handle this convection dominated regime, which is important for many applications, special numerical methods have been designed, such as the famous Streamline Upwind Petrov-Galerkin method [13]. They achieve stability by taking into account the direction in which quantities (such as fluid densities or momentum) are transported [32, 38]. This paper is devoted to how a form of upwinding, extending [1, 41, 12, 46], can be incorporated in an FES, expanding upon Example 5.31 in [20].

Section 3.2 serves as an appetizer, giving an introduction to the finite element method in dimension one, exemplifying problems and solutions related to vanishing viscosity. Section 3.3 gives some ingredients on discrete geometry useful for the definition of finite element systems, which is provided in Section 3.4. In Section 3.5 the main dish is served, namely the proposed upwinding technique for FES. Section 3.6 contains some remarks for further rumination.

3.2 Upwinding in dimension one

In this section we sketch a numerical method for solving one-dimensional convection-diffusion problems, which is adapted to the convection dominated regime. This well known method goes back at least to [1, 41] but is presented here in a finite element language, as in e.g. [33]. Our numerical illustrations follow [39] quite closely. This will give a quick introduction to the finite element method and motivate the generalizations presented in the following sections.

We choose $a < b$ in $\mathbb{R}$ and denote by I the interval $[a, b]$. For given $\alpha > 0$ (viscosity), $\beta \in \mathbb{R}$ (convection) and $f : I \to \mathbb{R}$ (source) we want to solve the second order differential equation for $u : I \to \mathbb{R}$:

$$-\alpha u'' + \beta u' = f, \tag{3.1}$$

$$u(a) = 0 \text{ and } u(b) = 0. \tag{3.2}$$

This problem can be given a variational formulation. Let X denote the space:

$$X = \mathrm{H}_0^1(I) = \{u \in \mathrm{L}^2(I) \ : \ u' \in \mathrm{L}^2(I), \ u(a) = 0, \ u(b) = 0\}.$$

The scalar product of $\mathrm{L}^2(I)$ and its extensions by continuity to dualities between Sobolev spaces will be denoted:

$$\langle u, v \rangle = \int uv.$$

We denote by A the differential operator:

$$A : u \mapsto -\alpha u'' + \beta u',$$

and define a bilinear form a on X by:

$$a(u,v) = \langle Au, v \rangle = \alpha \int u'v' + \int \beta u'v.$$

Introduce also a linear form l on X defined by:

$$l(v) = \langle f, v \rangle = \int fv.$$

The variational formulation of the above problem (3.1, 3.2) is:

$$u \in X, \ \forall v \in X \quad a(u,v) = l(v). \tag{3.3}$$

We now turn to the discretization of (3.3). Given a positive $n \in \mathbb{N}$ we choose points $x_i \in [a,b]$ for $i \in [\![0,n]\!]$, such that:

$$a = x_0 < \cdots < x_i < x_{i+1} < \cdots < x_n = b.$$

These points subdivide $[a,b]$ into n intervals of the form $[x_i, x_{i+1}]$ and we denote by h_n the length of the longest one:

$$h_n = \max\{|x_{i+1} - x_i| \ : \ i \in [\![0, n-1]\!]\}.$$

We let X_n denote the subspace of X consisting of functions that are piecewise affine with respect to this subdivision. The Galerkin discretization of (3.3) is:

$$u_n \in X_n, \ \forall v \in X_n \quad a(u_n, v) = l(v).$$

For the purposes of analysis, one considers a sequence of such subdivisions, indexed by n, providing a sequence (u_n) of approximate solutions to (3.3). For fixed $\alpha > 0$, $\beta \in \mathbb{R}$, $f \in \mathrm{L}^2(I)$ one has convergence of (u_n) to u, for instance in the sense:

$$\|u - u_n\|_{\mathrm{L}^2(I)} \leq C h_n^2 \text{ and } \|u' - u_n'\|_{\mathrm{L}^2(I)} \leq C' h_n.$$

Of course the constants C, C' depend on α, β, f. In practice we are interested in a regime where β and f are moderate ("of order 1"), whereas α is several magnitudes smaller. One analyses this regime by fixing $\beta \neq 0$ and f, and considering the asymptotic behavior of the sequence $u_n[\alpha]$, when $\alpha \to 0$ and $n \to \infty$. This is also referred to as the vanishing viscosity limit.

For illustration we chose $a = -1$, $b = 1$, $\beta = 1$ and $f = 1$. The exact solutions $u[\alpha]$ for various α are plotted in the top graph in Figure 3.1. As $\alpha \to 0$ the solution approaches the function $u[0] : I \to \mathbb{R}$ which solves:

$$\beta u' = f,$$
$$u(a) = 0.$$

Notice that of the two boundary conditions (3.2), only one is retained by the limit $u[0]$. At the other boundary point the small viscosity solutions $u[\alpha]$ display singular behavior, with formation of a boundary layer. Given that $\alpha > 0$, the part of the boundary where the homogeneous boundary condition is respected is the one where β, interpreted as a vector field on I, points into the domain, called the inflow boundary (a in our case). The boundary layer appears where β points out of the domain, called the outflow boundary (b in our case). The thickness of this layer is roughly α/β. Since the differential equation degenerates in its highest order term, this is an example of what is called a singular perturbation problem, for which the above (singular) behavior can be said to be typical.

For the Galerkin discretizations we have chosen equispaced points with $n = 25$. Results are plotted in the middle row of Figure 3.1. As the viscosity α approaches 0, the quality of the numerical solution deteriorates, with apparition of unwanted oscillations, extending far beyond the boundary layer. The numerical stability is often evaluated in terms of the dimensionless so-called Péclet number:

$$\text{Pé} = h_n \beta / \alpha.$$

In the numerical experiments, α was chosen in such a way that:

$$\text{Pé} = 10^k, \ k \in \{-0.5, 0, 0.5, 1\}.$$

The graphs illustrate that the numerical method rapidly deteriorates when Pé increases above 1. Notice that the problem is related to stability rather than consistency: at least away from the outflow boundary, the Galerkin space X_n contains rather good approximations of the exact solution, but the method is bad at choosing one.

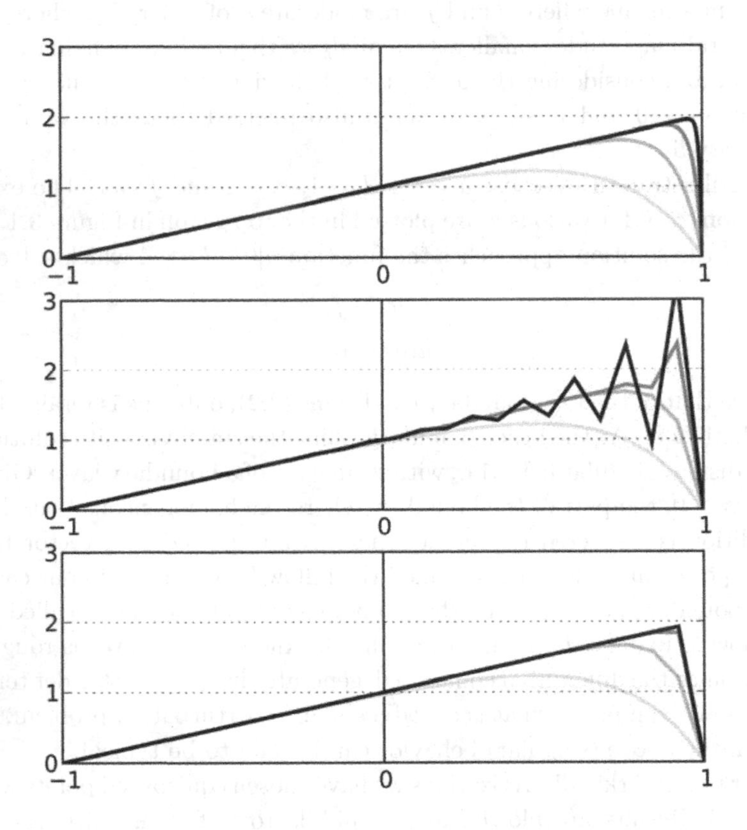

Figure 3.1 Decreasing viscosity from light gray to black. Exact solutions (top), standard piecewise affine finite elements (middle), affine trial functions and upwinded test functions (bottom).

A better numerical method can be obtained by cleverly choosing a new space Y_n and solving:

$$u_n \in X_n, \ \forall v \in Y_n \quad a(u_n, v) = l(v). \tag{3.4}$$

Such methods are called Petrov-Galerkin methods. One distinguishes between the trial space X_n and the test space Y_n. A rationale for the choice of Y_n can be obtained as follows. Choose a scalar product b on X. Let $a^\star$ be the transpose of a, the bilinear form associated with the adjoint differential operator:

$$A^\star : v \mapsto -\alpha v'' - \beta v'.$$

Denote by $\Phi : X \to X$ the map determined by solving the adjoint problems:

$$\forall v, w \in X \quad a^\star(\Phi(w), v) = b(w, v).$$

If we now solve (3.4) with $Y_n = \Phi(X_n)$ we notice the following. For all $w \in X_n$:

$$b(w, u_n) = a(u_n, \Phi(w)) = l(\Phi(w)) = a(u, \Phi(w)) = b(w, u).$$

Therefore u_n is the b-orthogonal projection of u on X_n. Orthogonal projection, which yields the best approximation with respect to the chosen norm, is typically a more stable procedure than the Galerkin projection associated with non-symmetric bilinear forms, such as a.

However this method may seem unpractical: for a given basis (e_i) of X_n it might seem too difficult a task to determine the basis $f_i = \Phi(e_i)$ of Y_n. Recall that the canonical basis of X_n consists of the functions $\lambda_i : I \to \mathbb{R}$, indexed by $i \in [\![1, n-1]\!]$, uniquely determined by the properties $\lambda_i \in X_n$ and $\lambda_i(x_j) = \delta_{ij}$ (Kronecker δ) for all $j \in [\![1, n-1]\!]$. As it turns out, for certain choices of b we can construct a similar basis for Y_n quite easily, providing not a free lunch but a bargain one.

Let b be defined by:

$$b(v, w) = \int v'w' = \langle Bv, w \rangle. \tag{3.5}$$

We notice that $B : v \mapsto -v''$ induces an isomorphism:

$$B : X_n \to Z_n = \bigoplus_{i \in [\![1, n-1]\!]} \mathbb{R}\, \delta_{x_i}, \tag{3.6}$$

with Dirac δ's at interior vertices. A basis $(\lambda_i^\star)$ of $Y_n = \Phi(X_n) = A^{\star -1}B(X_n)$ can be obtained by choosing $\lambda_i^\star : I \to \mathbb{R}$ to be the unique function satisfying $\lambda_i^\star(x_j) = \delta_{ij}$ for all $j \in [\![1, n-1]\!]$ (as for the standard basis) and :

$$A^\star \lambda_i^\star = 0 \text{ on }]x_j, x_{j+1}[\text{ for each } j \in [\![0, n-1]\!]. \tag{3.7}$$

Indeed $A^\star$ then induces an isomorphism:

$$A^\star : Y_n \to Z_n.$$

The basis functions $\lambda_i^\star$ can be determined explicitly. When $\beta = 0$, the

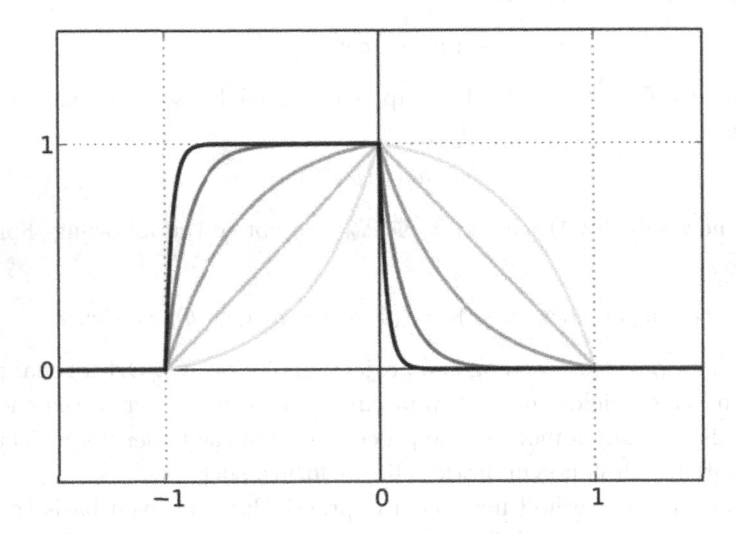

Figure 3.2 Basis functions, from moderately downwinded (light gray) to strongly upwinded (black). Wind blowing from left to right.

definition (3.7) yields the standard basis. For $\beta \neq 0$, $\lambda_i^\star$ is, on each sub-interval, a linear combination of a constant and an exponential:

$$\lambda_i^\star(x) = c_0 + c_1 \exp(rx) \text{ with } r = -\beta/\alpha, \text{ and } c_0, c_1 \in \mathbb{R}.$$

For $r = 0$ one should replace the above definition by affine functions. Typical basis functions are represented in Figure 3.2 for different values of r. Actually the chosen values are:

$$r = 10^k - 10^{-k} \text{ for } k \in \{0.5, 0, -0.5, -1, -1.5\}.$$

In the bottom row of Figure 3.1 the numerical solutions for the Petrov-Galerkin method (3.4) with upwinded basis functions for Y_n are plotted.

There are many senses in which, as $\alpha \to 0$, the solution $u[\alpha]$ converges to $u[0]$ — but the norm associated with (3.5) is not one of them. In fact:

$$\|u[\alpha]\|_b \to \infty \text{ as } \alpha \to 0,$$

precluding even weak convergence in X. This indicates that the b-norm is a poor choice for convergence analysis of the numerical method in the vanishing viscosity regime. From this point of view it is also a bit paradoxical that the numerical results are so satisfactory. A better but

α-dependent norm, is defined by:

$$b[\alpha](v, w) = \alpha \int v'w' + \int vw,$$

for which one can check:

$$\|u[\alpha] - u[0]\|_{b[\alpha]} \to 0 \text{ as } \alpha \to 0.$$

For the numerics we can use the approximation:

$$b[\alpha, n](v, w) = \alpha \int v'w' + \sum_i v(x_i)w(x_i)\mu_i,$$
$$\text{with } \mu_i = (x_{i+1} - x_{i-1})/2.$$

With this choice we still have an isomorphism (3.6) so we get the *same* numerical method as for (3.5).

Coincidentally, the choice $b[0, n]$ shows that the Petrov-Galerkin solution $u_n[\alpha]$ is in fact the element of X_n interpolating $u[\alpha]$ at vertices. We conclude that it is time to go to higher dimensions. The finite element method divides space (the interval) into cells, defines functions locally on these cells (with polynomials or exponentials for instance), imposing mere continuity between cells. The main point to remember, is that stability can be achieved by creating pairs of spaces where local functions match, through the differential operators of the problem to solve.

3.3 Discrete geometry

Useful references for the material sketched in this section include [25, 8].

3.3.1 Cellular complexes

For any natural number $k \geq 1$, let $\mathbb{B}^k$ be the closed unit ball of $\mathbb{R}^k$ and $\mathbb{S}^{k-1}$ its boundary. For instance $\mathbb{S}^0 = \{-1, 1\}$. We also put $\mathbb{B}^0 = \{0\}$.

Let S denote a compact metric space. A k-dimensional *cell* in S is a closed subset T of S for which there is a Lipschitz bijection $\mathbb{B}^k \to T$ with a Lipschitz inverse (if a cell T is both k- and l-dimensional then $k = l$). For $k \geq 1$, we denote by ∂T its boundary, the image of $\mathbb{S}^{k-1}$ by the chosen bi-Lipschitz map (different such maps give the same boundary). The interior of T is by definition $T \backslash \partial T$ (it is open in T but not necessarily in S).

Definition 3.1 A *cellular complex* is a pair $(S, \mathcal{T})$ where S is a compact metric space and $\mathcal{T}$ is a finite set of cells in S, such that the following conditions hold:

- Distinct cells in $\mathcal{T}$ have disjoint interiors.
- The boundary of any cell in $\mathcal{T}$ is a union of cells in $\mathcal{T}$.
- The union of all cells in $\mathcal{T}$ is S.

The subset of $\mathcal{T}$ consisting of k-dimensional cells is denoted $\mathcal{T}^k$.

$$\mathcal{T}^k = \{T \in \mathcal{T} \ : \ \dim T = k\}.$$

We also say that $\mathcal{T}$ is a cellular complex on S.

Example 3.2 Choose $a < b$ in $\mathbb{R}$ and let $S = [a, b]$. Choose points $x_i \in [a, b]$ for $i \in [\![0, n]\!]$, such that:

$$a = x_0 < \cdots < x_i < x_{i+1} < \cdots < x_n = b. \tag{3.8}$$

The following determines a cellular complex on S:

$$\begin{aligned} \mathcal{T}^0 &= \{x_i \ : \ i \in [\![0, n]\!]\}, \\ \mathcal{T}^1 &= \{[x_i, x_{i+1}] \ : \ i \in [\![0, n-1]\!]\}, \\ \mathcal{T} &= \mathcal{T}^0 \cup \mathcal{T}^1. \end{aligned}$$

In fact all cellular complexes on S are of this form, for a uniquely determined choice of points x_i subject to (3.8).

Example 3.3 Suppose we have two compact metric spaces M and N, equipped with cellular complexes $\mathcal{U}$ and $\mathcal{V}$. On $M \times N$ we let $\mathcal{U} \times \mathcal{V}$ denote the product cellular complex on $M \times N$, whose cells are all those of the form $U \times V$ for some $U \in \mathcal{U}$ and $V \in \mathcal{V}$.

This construction is illustrated in Figure 3.3 for the product of two intervals equipped with cellular complexes.

The boundary ∂T of any cell T of $\mathcal{T}$ can be naturally equipped with a cellular complex, namely:

$$\{T' \in \mathcal{T} \ : \ T' \subseteq T \text{ and } T' \neq T\}.$$

We use the same notation for the boundary of a cell and the cellular complex it carries.

A *refinement* of a cellular complex $\mathcal{T}$ on S is a cellular complex $\mathcal{T}'$ on S such that each element of $\mathcal{T}$ is the union of elements of $\mathcal{T}'$. We will be particularly interested in simplicial refinements of cellular complexes.

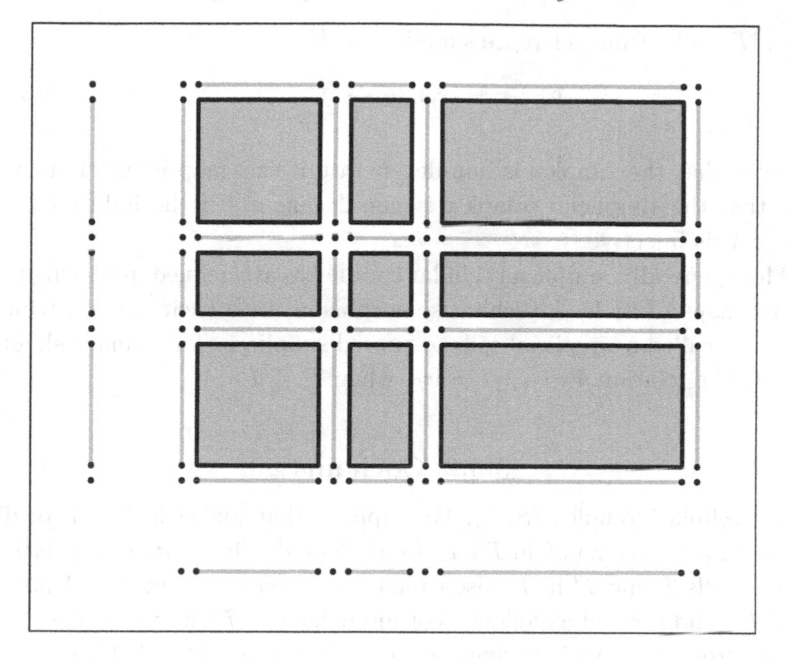

Figure 3.3 Product complex of two one-dimensional complexes. Cells are represented in gray, with black boundaries.

A cellular *sub*complex of a cellular complex $\mathcal{T}$ on S, is a cellular complex $\mathcal{T}'$ on some closed part S' of S such that $\mathcal{T}' \subseteq \mathcal{T}$. For instance if $T \in \mathcal{T}$ is a cell, its subcells form a subcomplex of $\mathcal{T}$, which we denote by $\widetilde{T}$. We have seen that the boundary of any cell $T \in \mathcal{T}$ can be equipped with a cellular complex which is a subcomplex of $\widetilde{T}$.

A *simplicial complex* on a set S, is a set $\mathcal{S}$ of non-empty *finite* subsets of S, such that for all T in $\mathcal{S}$ the non-empty subsets of T are also in $\mathcal{S}$. The elements of $\mathcal{S}$ are called simplices, the elements of a simplex vertices and the non-empty subsets of a simplex faces.

We attach a reference cell to any simplex $T \in \mathcal{S}$ as follows:

$$|T| = \{x = (x_i)_{i \in T} \in \mathbb{R}^T \ : \ \sum_{i \in T} x_i = 1 \text{ and } \forall i \in T \quad x_i \geq 0\}.$$

If $T' \subseteq T$ we have a canonical affine map $i_{TT'} : |T'| \to |T|$. The direct limit of this system is denoted $|\mathcal{S}|$. It is obtained from the disjoint union of reference simplices, identifying $|T'|$ with its image in $|T|$, in the preceding situation.

If S is a subset of an affine space V, there are natural maps

$\Phi_T : |T| \to V$, from reference simplices to V:

$$\Phi_T : x = (x_i)_{i \in T} \mapsto \sum_{i \in T} (x_i) i. \tag{3.9}$$

We say that the simplex is non-degenerate if this map is injective. We say that the simplicial complex is non-degenerate if the induced map $|\mathcal{S}| \to V$ is injective.

More generally, simplicial cellular complexes are defined, replacing the affine maps (3.9) by Lipschitz isomorphisms onto their ranges, which are then called a simplicial cells or curved simplices. These maps should satisfy the relation $\Phi_T \circ i_{TT'} = \Phi_{T'}$ when $T' \subseteq T$.

3.3.2 Cochains

Fix a cellular complex $(S, \mathcal{T})$. We suppose that for each $T \in \mathcal{T}$ of dimension ≥ 1, the manifold T has been oriented. The *relative orientation* of two cells T and T' in $\mathcal{T}$, also called the *incidence number*, is denoted $o(T, T')$ and defined as follows. For any edge $e \in \mathcal{T}^1$ its vertices are ordered, from say $\dot{e}$ to $\ddot{e}$. Define $o(e, \dot{e}) = -1$ and $o(e, \ddot{e}) = 1$. Concerning higher dimensional cells, fix $k \geq 1$. Given $T \in \mathcal{T}^{k+1}$ and $T' \in \mathcal{T}^k$ such that $T' \subseteq T$ we define $o(T, T') = 1$ if T' is outward oriented compared with T and $o(T, T') = -1$ if it is inward oriented. For all $T, T' \in \mathcal{T}$ not covered by these definitions we set $o(T, T') = 0$.

For each k, let $\mathcal{C}^k(\mathcal{T})$ denote the set of maps $c : \mathcal{T}^k \to \mathbb{R}$. Such maps associate a real number to each k-dimensional cell and are called *k-cochains.* The *coboundary* operator $\delta : \mathcal{C}^k(\mathcal{T}) \to \mathcal{C}^{k+1}(\mathcal{T})$ is defined by:

$$(\delta c)_T = \sum_{T' \in \mathcal{T}^k} o(T, T') c_{T'}.$$

The space of k-cochains has a canonical basis indexed by $\mathcal{T}^k$. The coboundary operator is the operator whose canonical matrix is the incidence matrix o, indexed by $\mathcal{T}^{k+1} \times \mathcal{T}^k$. A fundamental property of δ is that $\delta\delta = 0$ as a map $\mathcal{C}^k(\mathcal{T}) \to \mathcal{C}^{k+2}(\mathcal{T})$. In other words the family $\mathcal{C}^\bullet(\mathcal{T})$ is a complex. It is called the cochain complex and represented by:

$$0 \to \mathcal{C}^0(\mathcal{T}) \to \mathcal{C}^1(\mathcal{T}) \to \mathcal{C}^2(\mathcal{T}) \to \cdots$$

When S is a smooth manifold we denote by $\Omega^k(S)$ the space of smooth differential k-forms on S. Differential forms can be mapped to cochains as follows. Let S be a manifold and $\mathcal{T}$ a cellular complex on S. For each

k we denote by $\rho^k : \Omega^k(S) \to \mathcal{C}^k(\mathcal{T})$ the de Rham map, which is defined by:

$$\rho^k : u \mapsto \left(\int_T u \right)_{T \in \mathcal{T}^k}.$$

For each k, the following diagram commutes, as a consequence of Stokes theorem on each cell in $\mathcal{T}^{k+1}$:

$$\begin{array}{ccc} \Omega^k(S) & \xrightarrow{\mathrm{d}} & \Omega^{k+1}(S) \\ \downarrow \rho^k & & \downarrow \rho^{k+1} \\ \mathcal{C}^k(\mathcal{T}) & \xrightarrow{\delta} & \mathcal{C}^{k+1}(\mathcal{T}). \end{array}$$

A celebrated theorem of de Rham states that $\rho^\bullet$ induces isomorphisms on cohomology groups. Whitney forms, which we shall soon encounter, were introduced in [44, 45] (relative to smooth partitions of unity) as a tool of proof, but have recently reappeared in numerical analysis (relative to piecewise affine partitions of unity). The next section introduces a general framework containing these forms as well as useful generalizations.

3.4 Finite element systems

3.4.1 Definition

For the purposes of analysis of partial differential equations and numerical methods, smooth differential forms are somewhat insufficient. Banach spaces of differential forms with various integrability conditions are used. For linear equations, L^2 based Hilbert spaces are often enough.

For any cell T, we denote by $X^k(T)$ the space of differential k-forms on T which are $\mathrm{L}^2(T)$ and have exterior derivative, a priori defined as a current, in $\mathrm{L}^2(T)$. We let $X^k_0(T)$ be the closure of the subspace of $X^k(T)$ consisting of forms whose support is a compact subset of the interior of T. If T' is a face of T, let $i_{TT'} : T' \to T$ be the inclusion map. When T' has codimension 1 in T, the pullback of k-forms by $i_{TT'}$ defines an operator which is bounded from $X^k(T)$ to the dual of $X_0^{\dim T' - k}(T')$. It therefore makes sense to define:

$$Y^k(T) = \{u \in X^k(T) \ : \forall T' \in \mathcal{T} \quad T' \subseteq T \Rightarrow u|_{T'} \in X^k(T')\}.$$

Definition 3.4 Suppose $\mathcal{T}$ is a cellular complex. For each $k \in \mathbb{N}$ and each $T \in \mathcal{T}$ we suppose we are given a space $E^k(T) \subseteq Y^k(T)$, called a differential k-element on T. The following conditions should be satisfied:

- If $i_{TT'} : T' \subseteq T$ is an inclusion of cells, pullback induces a map $i^\star_{TT'} : E^k(T) \to E^k(T')$.
- The exterior derivative induces maps $\mathrm{d} : E^k(T) \to E^{k+1}(T)$, for all cells $T \in \mathcal{T}$ and all k.

Such a family of elements is called an *element system.*

A differential element is said to be finite if it is finite dimensional. A finite element system, abbreviated FES in the following, is an element system in which all the elements are finite.

For any subcomplex $\mathcal{T}'$ of $\mathcal{T}$ we define $E^k(\mathcal{T}')$ as follows :

$$E^k(\mathcal{T}') = \{u \in \bigoplus_{T \in \mathcal{T}'} E^k(T) \ : \ \forall T, T' \in \mathcal{T} \quad T' \subseteq T \Rightarrow u_T|_{T'} = u_{T'}\}.$$

An FES over a cellular complex is an *inverse system* of complexes. To inclusion of cells $T' \subseteq T$, correspond pullback operators $i^\star_{TT'}$, which are morphisms of complexes with some obvious properties:

$$i^\star_{TT} = \mathrm{id},$$
$$i^\star_{T'T''} \circ i^\star_{TT'} = i^\star_{TT''}, \text{ when } T'' \subseteq T' \subseteq T.$$

The space $E^\bullet(\mathcal{T}')$ defined above is an *inverse limit* of this system and is determined by this property up to unique isomorphism. For a cell $T \in \mathcal{T}$, its collection of subcells is the cellular complex $\widetilde{T}$. Applied to this case, the above definition of $E^k(\widetilde{T})$ gives a space canonically isomorphic to $E^k(T)$, so that we can identify $E^k(\widetilde{T}) = E^k(T)$.

Elements of $E^k(\mathcal{T}')$ may be regarded as differential forms defined cellwise, which have a partial continuity property across interfaces between cells, in the form of equal pullbacks.

Applied to $\mathcal{T}$, this will provide spaces $E^k(\mathcal{T})$ suitable for Galerkin and Petrov-Galerkin discretizations of PDEs expressed in terms of exterior derivatives.

Also of importance is the application of the above construction to the boundary ∂T of a cell T, considered as a cellular complex consisting of all subcells of T except T itself. Considering ∂T as a cellular complex (not only a subset of T) we denote the constructed inverse limit by $E^k(\partial T)$. If $i : \partial T \to T$ denotes the inclusion map, the pullback by i defines a map $i^\star : E^k(T) \to E^k(\partial T)$ which we denote by ∂ and call restriction.

Given an element system E on a cellular complex $\mathcal{T}$, we consider now the following conditions:

- *Extensions.* For each $T \in \mathcal{T}$ and $k \in \mathbb{N}$, the restriction $\partial : E^k(T) \to E^k(\partial T)$ is onto.
- *Exactness.* For each $T \in \mathcal{T}$ the following sequence, concerning the exterior derivative, is exact:

$$0 \to \mathbb{R} \to E^0(T) \to E^1(T) \to \cdots \to E^{\dim T}(T) \to 0.$$

Definition 3.5 We will say that an element system is *compatible* if the two conditions above hold.

The first surjectivity condition can be written symbolically $\partial E^k(T) = E^k(\partial T)$. We denote also:

$$\begin{aligned} E_0^k(T) &= \ker \partial : E^k(T) \to E^k(\partial T), \\ &= E^k(T) \cap X_0^k(T). \end{aligned}$$

The latter equality requires some work.

The following is Theorem 3.1 of [16], see also Proposition 5.16 in [20], and is comparable to well-known so called piecewise de Rham theorems, see e.g. [25].

Proposition 3.6 *If E is a compatible finite element system, the de Rham map $\rho^\bullet : E^\bullet(\mathcal{T}) \to \mathcal{C}^\bullet(\mathcal{T})$ induces isomorphisms on cohomology groups.*

Compatible FES also have fairly local bases, see Remark 3.17.

3.4.2 Examples

We now provide some examples of element systems, *all of which are compatible.*

Example 3.7 The spaces $Y^k(T)$ themselves define an element system. It is far from finite.

On the other hand the spaces $X^k(T)$ do not constitute an element system. This family of spaces is stable under the exterior derivative, but not under restriction from cells to subcells.

Example 3.8 In dimension 1, consider a cellular complex with notation as in Example 3.2. In the following examples $E^0(\{x_i\}) = \mathbb{R}^{\{x_i\}} \approx \mathbb{R}$ for each i.

- *Piecewise polynomials.* The space of polynomials of degree at most p is denoted $\mathbb{P}_p$. Pick an integer $p_i \geq 1$ for each $i \in [\![0, n-1]\!]$. Define:

$$E^0([x_i, x_i + 1]) = \mathbb{P}_{p_i},$$
$$E^1([x_i, x_i + 1]) = \mathbb{P}_{p_i - 1}\mathrm{d}x.$$

Here, restriction of the polynomials to the interval is implicit.

- *Piecewise exponentials.* Pick a real $r_i \neq 0$ for each $i \in [\![0, n-1]\!]$. Define:

$$E^0([x_i, x_i + 1]) = \{x \mapsto c_0 + c_1 \exp(r_i x) \ : \ c_0, c_1 \in \mathbb{R}\},$$
$$E^1([x_i, x_i + 1]) = \{x \mapsto c \exp(r_i x)\mathrm{d}x \ : \ c \in \mathbb{R}\}.$$

- *Splines*, see [43], can also be accommodated. Inserting knots in cells $[x_i, x_{i+1}]$ is fine, but between cells one has mere continuity in E^0, and no continuity in E^1.
- *More generally* take, on each interval, for E^0, a space of functions containing at least a constant function and one with different values at the extremities, and for E^1 the space of derivatives.

Example 3.9 In the situation of Example 3.3, suppose that one has differential elements $E^k(U)$ for $U \in \mathcal{U}$ and $F^k(V)$ for $V \in \mathcal{V}$, both forming systems according to Definition 3.4. It is then possible to define a natural notion of tensor product of E and F, as an FES on the product complex of $\mathcal{U}$ and $\mathcal{V}$, see [20] p. 77–79. As one might expect, if E and F are compatible, in the sense of Definition 3.5, then so is their tensor product.

This can be combined with the preceding Examples to define FES on products of intervals.

Example 3.10 Let V be a finite dimensional vector space. We denote by $\mathbb{P}_p(V)$ the space of polynomials on V of degree at most p, by $\mathbb{A}^k(V)$ the space of alternating k-linear maps $V^k \to \mathbb{R}$ and by $\mathbb{PA}^k_p(V)$ the space of differential k-forms on V which are polynomials of degree at most p.

Denote the Koszul operator by κ. It is the contraction of differential forms by the identity, interpreted as a vector field on V:

$$(\kappa u)_x(\xi_1, \ldots, \xi_k) = u_x(x, \xi_1, \ldots, \xi_k).$$

Alternatively one can use the Poincaré operator associated with the canonical homotopy from the identity to the null-map. Let $\mathcal{T}$ be a simplicial complex. Define, for any non-zero $p \in \mathbb{N}$, any simplex $T \in \mathcal{T}$ and any $k \in \mathbb{N}$:

$$\begin{aligned}\Lambda^k_p(T) &= \{u \in \mathbb{PA}^k_p(T) \ : \ \kappa u \in \mathbb{PA}^{k-1}_p(T)\}, \qquad (3.10)\\ &= \mathbb{PA}^k_{p-1}(T) + \kappa \mathbb{PA}^{k+1}_{p-1}(T).\end{aligned}$$

For fixed p, we call this the *trimmed* polynomial finite element system of order p.

This construction, due to [36] for $\mathbb{R}^2$, [34] for vector fields in $\mathbb{R}^3$, and [27] for differential forms, is thoroughly treated in [3]. The case $p = 1$ corresponds to constructs in [44, 45], called Whitney forms, as pointed out in [7].

Whitney forms are defined as follows. For a given simplicial mesh $\mathcal{T}$, let λ_i denote the barycentric coordinate map associated with vertex i. It is the unique continuous and piecewise affine function taking the value 1 at vertex i and the value 0 at other vertices $j \in \mathcal{T}^0$. To a simplex $T \in \mathcal{T}^k$ of dimension $k \geq 1$ one attaches the k-form λ_T defined by:

$$\lambda_T = k! \sum_{i=0}^{k} (-1)^i \lambda_i \mathrm{d}\lambda_0 \wedge \cdots (\mathrm{d}\lambda_i)^\wedge \cdots \wedge \mathrm{d}\lambda_k. \qquad (3.11)$$

Here, the vertices of T are labeled $\{0, \ldots, k\}$, respecting the orientation of T. The superscript $(\cdot)^\wedge$ signifies omission of this term. The λ_T for $T \in \mathcal{T}^k$ constitute a basis for $\Lambda^k_1(\mathcal{T})$.

It was usual to start the indexing in (3.10) at $p = 0$ but, as remarked in the preprint of [15], the advantage of letting the lowest order be $p = 1$ is that the wedge product induces maps:

$$\wedge : \Lambda^{k_0}_{p_0}(T) \times \Lambda^{k_1}_{p_1}(T) \to \Lambda^{k_0+k_1}_{p_0+p_1}(T).$$

In words, the wedge product respects the grading in k and the filtering in p.

Example 3.11 A convenient way of constructing a compatible FES on a cellular complex $\mathcal{T}$ is to take a refinement $\mathcal{T}'$ on which one has a compatible FES $E^k(T)$, and consider this as an FES on $\mathcal{T}$. Namely, if $T \in \mathcal{T}$, consider T as a cellular subcomplex $\widetilde{T}$ of $\mathcal{T}'$ and define $F^k(T) = E^k(\widetilde{T})$. The Galerkin spaces associated with E and F are identical, even though they are different as FES. The versatility of such *composite* finite elements has already been put to good use, perhaps most famously in the Clough-Tocher element of class $C^1(S)$.

Example 3.12 For an interpretation of so-called *hp* finite elements in this framework, see Example 5.24, p. 79 in [20].

Remark 3.13 This point of view on finite elements is at variance with Ciarlet's definition, as provided, for instance, in [22] §10.

- Firstly, we consider fields not just on cells of maximal dimension but also on all their faces.
- Secondly, we consider at once differential forms of all degrees, rather that say just functions or just vector fields.
- Thirdly, so called degrees of freedom are not part of the definition, instead inter-element continuity is enforced by conditions on pullbacks to faces.

Our definition is more restrictive in the sense that we have not accommodated say finite elements for matrix fields such as those of elasticity or those with higher order inter-element continuity such as Hermite polynomials. On the other hand it is tailored to account for the successes of the elements defined in Example 3.10.

3.5 Upwinding in FES

3.5.1 Locally harmonic forms

Let T be a cell where, for each k, $E^k(T)$ is equipped with a scalar product b. Orthogonality with respect to b will be denoted $\perp$. We say that a k-form u on T is *E–harmonic* if:

$$\mathrm{d}u \perp \mathrm{d}E_0^k(T) \text{ and } u \perp \mathrm{d}E_0^{k-1}(T). \tag{3.12}$$

One can for instance take b to be the $\mathrm{L}^2(T)$ scalar product on differential forms, associated with some Riemannian metric. Denote by $\mathrm{d}^\star$ the formal adjoint of d with respect to this scalar product. The continuous analogue of the above condition (3.12) is:

$$\mathrm{d}^\star \mathrm{d}u = 0 \text{ and } \mathrm{d}^\star u = 0. \tag{3.13}$$

From the other point of view, (3.12) is the Galerkin variant of (3.13).

Let E be a finite element system on $\mathcal{T}$. Define a finite element system $\mathring{E}$ by:

$$\mathring{E}^k(T) = \{u \in E^k(T) \ : \ \forall T' \in \mathcal{T} \quad T' \subseteq T \ \Rightarrow \ u|_{T'} \text{ is } E\text{–harmonic}\}.$$

We say that $\mathring{E}$ is the subsystem of E of locally harmonic forms.

The following was essentially proved in [16].

Proposition 3.14 *If E is a compatible FE system then $\mathring{E}$ is a compatible FE system such that the de Rham maps $\rho^k : \mathring{E}^k(\mathcal{T}) \to \mathcal{C}^k(\mathcal{T})$ are isomorphisms.*

Example 3.15 This construction generalizes [30], in which div-conforming finite element vector fields are defined on polyhedral meshes in $\mathbb{R}^3$.

Remark 3.16 Since $\mathcal{C}^k(\mathcal{T})$ has a canonical basis, the de Rham map determines a corresponding canonical basis of $\mathring{E}^k(\mathcal{T})$. Its elements can be constructed by recursive harmonic extension. Explicitly, to construct the k-form u_T attached to a given $T \in \mathcal{T}^k$, define $u_T|_T$ to be the unique element of $E^k(T)$ orthogonal to $\mathrm{d}E_0^{k-1}(T)$ and with integral 1. For $T' \in \mathcal{T}^k$, different from T, set $u_T|_{T'} = 0$. Supposing u has been defined on the l-skeleton of $\mathcal{T}$ (constituted by cells of dimension at most l), u is extended to the $(l+1)$-skeleton through harmonic extension to $(l+1)$-dimensional cells, as defined by (3.12).

Remark 3.17 More generally it is possible, for any compatible FES E, to define isomorphisms:

$$\bigoplus_{T \in \mathcal{T}} E_0^k(T) \to E^k(\mathcal{T}),$$

through a procedure of recursive harmonic extension, see Proposition 3.14 in [17] (also Proposition 3.1 and Remark 3.3 in [16]). More general isomorphisms are constructed in [4].

Example 3.18 Start with a fine cellular complex, and coarsen it by agglomerating cells, as in Example 3.11. An FES on the fine mesh yields an FES on the coarse mesh, of which one can consider the subsystem of locally harmonic forms. The construction can be applied recursively, agglomerating cells at each step, and taking at a given level the locally harmonic fields of the refined level. This yields nested spaces that can be used for multilevel preconditioning [35].

Example 3.19 On simplices, lowest order trimmed polynomial differential forms are locally harmonic in the sense of (3.13), with respect to the $\mathrm{L}^2(T)$ product associated with *any* Euclidean metric (a Riemannian metric which is constant on T), as pointed out in §4.1 of [16]. Whitney forms, as defined by (3.11), provide an explicit solution to the problem of recursive harmonic extension.

Example 3.20 For a given complex $\mathcal{T}$, consider its barycentric refinement $\mathcal{T}'$ (assuming it is well defined), and reassemble the simplices of $\mathcal{T}'$

to form the dual $\mathcal{T}^\star$ of $\mathcal{T}$. Take lowest order trimmed polynomial differential forms on $\mathcal{T}'$. Consider them once as an FES on $\mathcal{T}$ and once as an FES on $\mathcal{T}^\star$, as in Example 3.11. For these two FES consider their subsystems of locally harmonic forms. At least in some cases, one can prove that these two subsystems are in duality, providing a finite element analogue of Hodge duality, which has applications to preconditioning [14]. For a tentative application to fluid dynamics, see Example 5.30 in [20].

Remark 3.21 Local harmonicity appears to behave naturally with respect to the tensor products of Example 3.9.

3.5.2 Application to upwinding

We now apply the construction of locally harmonic forms to obtain a generalization of Section 3.2 to multidimensional problems, extending also, for instance, [12]. We have in mind applications to fluid dynamics. A model problem for this situation is the following. Consider a bounded domain S in a Euclidean space, $\alpha > 0$, a divergence free vector field V on S and also a function f on S. Consider the equation for a scalar field u on S:

$$-\alpha \Delta u + V \cdot \operatorname{grad} u = f \text{ in } S, \tag{3.14}$$

$$u|_{\partial S} = 0. \tag{3.15}$$

The variational form of this equation is:

$$u \in \mathrm{H}^1_0(S),\ \forall v \in \mathrm{H}^1_0(S) \quad \int \operatorname{grad} u \cdot (\alpha \operatorname{grad} v + V v) = \int f v.$$

One is interested in the asymptotic behavior of u for small positive α. One is also interested in similar problems for vector fields. Typically the flow field itself satisfies an equation of this type. More generally, analogous equations for differential forms are of interest. We are going to design Petrov-Galerkin methods of the form (3.4) for this purpose.

Consider a cellular complex and a large compatible finite element system on it, obtained for instance by considering trimmed polynomial differential forms on a simplicial refinement, as in Example 3.10. We let one space (say the trial space) consist of locally harmonic forms for the standard L^2 product, deduced from the Euclidean metric. For the other space (say the test space) we use locally harmonic forms for a weighted L^2 product. To motivate our choice of weight we introduce some machinery.

Let β be a one-form on S. The associated covariant exterior derivative is:

$$\mathrm{d}_\beta : u \mapsto \mathrm{d}u + \beta \wedge u. \tag{3.16}$$

These operators do not form a complex but we have:

$$\mathrm{d}_\beta \mathrm{d}_\beta u = (\mathrm{d}\beta) \wedge u.$$

In gauge theory (see [29]) the term $\mathrm{d}\beta$ is called curvature. Supposing that $\beta = \mathrm{d}\phi$ for a function ϕ we have:

$$\mathrm{d}_\beta u = \exp(-\phi)\mathrm{d}\exp(\phi)u.$$

One says that $u \mapsto \exp(\phi)u$ is a gauge transformation.

We denote by $\mathrm{d}_\beta^\star$ the formal adjoint of d_β with respect to the L^2 scalar product on differential forms. When $\beta = \mathrm{d}\phi$ we have:

$$\mathrm{d}_\beta^\star u = \exp(\phi)\mathrm{d}^\star \exp(-\phi)u.$$

A natural generalization of (3.13) is then:

$$\mathrm{d}_\beta^\star \mathrm{d}u = 0 \text{ and } \mathrm{d}_\beta^\star u = 0,$$

which can be written:

$$\mathrm{d}^\star \exp(-\phi)\mathrm{d}u = 0 \text{ and } \mathrm{d}^\star \exp(-\phi)u = 0. \tag{3.17}$$

Returning to (3.14, 3.15) we suppose S is a domain in a Euclidean space. We fix a cellular complex $\mathcal{T}$ on S, with flat cells, and a large compatible FE system E on it, as indicated above. This means that $E_0^k(T)$ should have large enough dimension, to accommodate a variety of differential forms. We construct two spaces of locally harmonic forms deduced from E, distinguished by the choice of scalar product b defining orthogonality in (3.12). For one (the trial space) take b to be the standard L^2 scalar product. For the other (the test space) we introduce the 1-form β defined by:

$$\beta_x(\xi) = V(x) \cdot \xi.$$

We choose for each $T \in \mathcal{T}$, a constant approximation β_T of the pullback of β to T. Let ϕ_T be an affine function such that $\mathrm{d}\phi_T = \beta_T$. One can determine it completely for instance by imposing zero mean on T. To mimic (3.17) we use the scalar product:

$$b(u, v) = \int_T \exp(\phi_T/\alpha)u \cdot v,$$

to define the locally harmonic functions. For explicitness, recall that the local trial functions $u \in \mathring{E}^k(T)$ should then satisfy the equations:

$$\forall v \in E_0^k(T) \quad \int \exp(\phi_T/\alpha) \mathrm{d}u \cdot \mathrm{d}v = 0, \tag{3.18}$$

$$\forall v \in E_0^{k-1}(T) \quad \int \exp(\phi_T/\alpha) u \cdot \mathrm{d}v = 0. \tag{3.19}$$

The canonical basis of the test space, as defined by Remark 3.16, will then be upwinded, compared with the canonical basis of the trial space. To get an idea of the shape of upwinded basis functions we can remark that when v is a constant differential k-form on T, $u = \exp(-\phi_T/\alpha)v$ satisfies the equations (3.18, 3.19), whatever element system is used.

Example 3.22 In dimension one we recover the upwinded basis functions defined in Section 3.2 from the infinite dimensional element system Y. These are "true" locally harmonic functions, with respect to an exponential weight. In higher dimensions one cannot hope to have explicit formulas for these, which is why we have introduced discrete analogues. The challenge was to get a discrete analogue where basic properties, such as local sequence exactness with respect to the exterior derivative, are satisfied.

Remark 3.23 There is at least one case in which one can get an explicit formula for the upwinded differential forms, namely the extension of 0-forms from vertices to the 1-skeleton. In some cases this is all one needs. Let $\mathcal{T}$ be a simplicial complex. Define a coefficient μ_e attached to the edge $e \in \mathcal{T}^1$ with vertices i and j, by:

$$\mu_e = -\int_S \mathrm{d}\lambda_i \cdot \mathrm{d}\lambda_j.$$

We suppose the simplices are such that $\mu_e \geq 0$. For Whitney 1-forms $u, v \in \Lambda_1^1(\mathcal{T})$ one then has an estimate of the form:

$$\left| \int u \cdot v - \sum_e \mu_e (\rho^1 u)_e (\rho^1 v)_e \right| \leq Ch \left(\|u\|_{\mathrm{L}^2} \|\mathrm{d}v\|_{\mathrm{L}^2} + \|\mathrm{d}u\|_{\mathrm{L}^2} \|v\|_{\mathrm{L}^2} \right),$$

where we sum over edges $e \in \mathcal{T}^1$ and h represents the largest diameter of a simplex of $\mathcal{T}$, see [26] (or [19]). This indicates that the discrete sum on the left hand side is a good approximation of the integral.

Define on $\mathrm{H}_0^1(S) \times \mathrm{H}_0^1(S)$:

$$a(u, v) = \int \operatorname{grad} u \cdot (\alpha \operatorname{grad} v + Vv),$$

and its approximation (ρ is the de Rham map):

$$a_h(u,v) = \sum_e \mu_e(\rho^1 \mathrm{d}u)_e(\rho^1(\alpha \mathrm{d}v + \beta_e v))_e.$$

Letting $u \in \Lambda^0_1(\mathcal{T})$ and v be defined at vertices and upwinded on edges according to (3.18) with $E = Y$, we recover the scheme of [46].

Remark 3.24 Residual Free Bubbles (see e.g. [10]) on the other hand, from which the Streamline Upwind Petrov-Galerkin method [13] can be deduced, work by extending standard finite elements on cells of maximal dimension only, by the whole space $X^k_0(T)$, and reducing the resulting equation to a modified one on the original space in special circumstances.

3.6 Further topics

High order upwinding. Let E be a large compatible FES, providing workspace, and let F be a compatible FES that we wish to upwind, within E. For any function ϕ defined on the simplices of $\mathcal{T}$, we denote by b_ϕ the weighted L^2 product of forms on T:

$$b_\phi(u,v) = \int_T \exp(\phi_T) u \cdot v.$$

We define:

$$\begin{aligned} G^k(T) = \{u \in E^k(T) \ : \ &\forall T' \subseteq T \ \exists v \in F^k_0(T') \\ &\forall w \in E^k_0(T') \quad b_\phi(\mathrm{d}u, \mathrm{d}w) = b_0(\mathrm{d}v, \mathrm{d}(\exp(\phi)w)), \\ &\forall w \in E^{k-1}_0(T') \quad b_\phi(u, \mathrm{d}w) = b_0(v, \mathrm{d}(\exp(\phi)w))\}. \end{aligned}$$

Then G appears to define a compatible FES which has the same dimension as F and is a locally upwinded version of it (scaling ϕ by α).

Minimal FES. The framework of FES can also be applied to construct a *minimal compatible* FES containing certain prespecified differential forms, see [18]. The trimmed polynomial finite element system of order p is a minimal compatible one containing polynomial differential forms of order $p - 1$.

Vector bundles. The covariant exterior derivative defined by (3.16) corresponds to a very special choice of vector bundle, namely one with fiber $\mathbb{R}$ and gauge group $\mathbb{R}_+$ consisting of the positive reals under multiplication. A somewhat more elaborate variant uses fibers $\mathbb{C}$ and gauge group $\mathbb{U}(1)$ consisting of the complex numbers with modulus 1. For an extension of the present theory to this case, with applications to wave equations of Helmholtz type, see Example 5.32 in [20].

Analysis. At present we are quite far from a satisfactory analysis of upwinded FES. Even for one dimensional problems there are recent interesting developments, e.g. [40]. For standard FES, such as trimmed polynomial differential forms (Example 3.10), a powerful tool for convergence proofs is the construction of projections that commute with the exterior derivative, yet are uniformly bounded with respect to L^2 norms, [42, 15, 3, 21]. The important property of these examples is that they are scale invariant FES, and the analysis has been carried over to this setting in [20] §5.2–4. Various L^p estimates are obtained, yielding discrete Sobolev injections and translation estimates. However, upwinded FES are not scale invariant, since they take into account certain length scales, such as the thickness of the boundary layers.

References

[1] D. N. de G. Allen and R. V. Southwell, Relaxation methods applied to determine the motion, in two dimensions, of a viscous fluid past a fixed cylinder. *Quart. J. Mech. Appl. Math.*, **8**, (1955), 129–145.

[2] D. N. Arnold, Differential complexes and numerical stability. In *Proceedings of the International Congress of Mathematicians, Vol. I (Beijing, 2002)*, 137–157, Beijing, 2002, Higher Ed. Press.

[3] D. N. Arnold, R. S. Falk and R. Winther, Finite element exterior calculus, homological techniques, and applications. *Acta Numer.*, **15**, (2006), 1–155.

[4] D. N. Arnold, R. S. Falk and R. Winther, Geometric decompositions and local bases for spaces of finite element differential forms. *Comput. Methods Appl. Mech. Engrg.*, **198**, (2009), 1660–1672.

[5] D. N. Arnold, R. S. Falk and R. Winther, Finite element exterior calculus: from Hodge theory to numerical stability. *Bull. Amer. Math. Soc. (N.S.)*, **47**, (2010), 281–354.

[6] P. B. Bochev and J. M. Hyman, Principles of mimetic discretizations of differential operators. In *Compatible Spatial Discretizations*, **142** of *IMA Vol. Math. Appl.*, 89–119. Springer, New York, 2006.

[7] A. Bossavit, Mixed finite elements and the complex of Whitney forms. In *The Mathematics of Finite Elements and Applications, VI (Uxbridge, 1987)*, 137–144. Academic Press, London, 1988.

[8] R. Bott and L. W. Tu, *Differential Forms in Algebraic Topology*. **82** of *Graduate Texts in Mathematics*, Springer-Verlag, New York, 1982.

[9] F. Brezzi and M. Fortin, *Mixed and Hybrid Finite Element Methods*, **15** of *Springer Series in Computational Mathematics*, Springer-Verlag, New York, 1991.

[10] F. Brezzi, T. J. R. Hughes, L. D. Marini, A. Russo and E. Süli, A priori error analysis of residual-free bubbles for advection-diffusion problems. *SIAM J. Numer. Anal.*, **36**, (1999), 1933–1948.

[11] F. Brezzi, K. Lipnikov and M. Shashkov, Convergence of the mimetic finite difference method for diffusion problems on polyhedral meshes. *SIAM J. Numer. Anal.*, **43**, (2005), 1872–1896.

[12] F. Brezzi, L. D. Marini and P. Pietra, Two-dimensional exponential fitting and applications to drift-diffusion models. *SIAM J. Numer. Anal.*, **26**, (1989), 1342–1355.

[13] A. N. Brooks and T. J. R. Hughes, Streamline upwind/Petrov-Galerkin formulations for convection dominated flows with particular emphasis on the incompressible Navier-Stokes equations. *Comput. Methods Appl. Mech. Engrg.*, **32** (1982), 199–259.

[14] A. Buffa and S. H. Christiansen, A dual finite element complex on the barycentric refinement. *Math. Comp.*, **76**, (2007), 1743–1769.

[15] S. H. Christiansen, Stability of Hodge decompositions in finite element spaces of differential forms in arbitrary dimension. *Numer. Math.*, **107**, (2007), 87–106.

[16] S. H. Christiansen, A construction of spaces of compatible differential forms on cellular complexes. *Math. Models Methods Appl. Sci.*, **18**, (2008), 739–75.

[17] S. H. Christiansen, Foundations of finite element methods for wave equations of Maxwell type. In *Applied Wave Mathematics*, 335–393. Springer, Berlin Heidelberg, 2009.

[18] S. H. Christiansen, Éléments finis mixtes minimaux sur les polyèdres. *C. R. Math. Acad. Sci. Paris*, **348**, (2010), 217–221.

[19] S. H. Christiansen and T. G. Halvorsen, A gauge invariant discretization on simplicial grids of the Schrödinger eigenvalue problem in an electromagnetic field. *SIAM J. Numer. Anal.*, **49**, (2011), 331–345.

[20] S. H. Christiansen, H. Z. Munthe-Kaas and B. Owren, Topics in structure-preserving discretization. *Acta Numer.*, **20**, (2011), 1–119.

[21] S. H. Christiansen and R. Winther, Smoothed projections in finite element exterior calculus. *Math. Comp.*, **77**, (2008), 813–829.

[22] P. G. Ciarlet, Basic error estimates for elliptic problems. In *Handbook of Numerical Analysis, Vol. II*, 17–351, North-Holland, Amsterdam, 1991.

[23] J. Dodziuk and V. K. Patodi, Riemannian structures and triangulations of manifolds. *J. Indian Math. Soc. (N.S.)*, **40**, (1977), 1–52.

[24] J. Droniou, R. Eymard, T. Gallouët and R. Herbin, A unified approach to mimetic finite difference, hybrid finite volume and mixed finite volume methods. *Math. Models Methods Appl. Sci.*, **20**, (2010), 265–295.

[25] P. A. Griffiths and J. W. Morgan, *Rational Homotopy Theory and Differential Forms*. In *Progress in Mathematics*, **16**, Birkhäuser Boston, Mass., 1981.

[26] Y. Haugazeau and P. Lacoste, Condensation de la matrice masse pour les éléments finis mixtes de $H(\mathrm{rot})$. *C. R. Acad. Sci. Paris*, **316**, (1993), 509–512.

[27] R. Hiptmair, Canonical construction of finite elements. *Math. Comp.*, **68**, (1999), 1325–1346.

[28] R. Hiptmair, Finite elements in computational electromagnetism. *Acta Numer.*, **11**, (2002), 237–339.

[29] S. Kobayashi and K. Nomizu, *Foundations of Differential Geometry. Vol I.* Interscience Publishers, New York–London, 1963.

[30] Y. Kuznetsov and S. Repin, Convergence analysis and error estimates for mixed finite element method on distorted meshes. *J. Numer. Math.*, **13**, (2005), 33–51.

[31] P. Monk, *Finite Element Methods for Maxwell's Equations.* Oxford University Press, New York, 2003.

[32] K. W. Morton, *Numerical Solution of Convection-Diffusion Problems.* In *Applied Mathematics and Mathematical Computation*, **12**, Chapman & Hall, London, 1996.

[33] K. W. Morton, The convection-diffusion Petrov-Galerkin story. *IMA J. Numer. Anal.*, **30**, (2010), 231–240.

[34] J.-C. Nédélec, Mixed finite elements in $\mathbf{R}^3$. *Numer. Math.*, **35**, (1980), 315–341.

[35] J. E. Pasciak and P. S. Vassilevski, Exact de Rham sequences of spaces defined on macro-elements in two and three spatial dimensions. *SIAM J. Sci. Comput.*, **30**, (2008), 2427–2446.

[36] P.-A. Raviart and J. M. Thomas, A mixed finite element method for 2nd order elliptic problems. In *Mathematical Aspects of Finite Element Methods (Proc. Conf., Consiglio Naz. delle Ricerche (C.N.R.), Rome, 1975)*, 292–315. Lecture Notes in Math., **606**, Springer, Berlin, 1977.

[37] J. E. Roberts and J.-M. Thomas, Mixed and hybrid methods. In *Handbook of Numerical Analysis, Vol. II*, 523–639, North-Holland, Amsterdam, 1991.

[38] H.-G. Roos, M. Stynes and L. Tobiska, *Robust Numerical Methods for Singularly Perturbed Differential Equations.* In *Springer Series in Computational Mathematics*, **24**, Springer-Verlag, 2nd edition, 2008.

[39] A. Russo, Streamline-upwind Petrov/Galerkin method (SUPG) vs residual-free bubbles (RFB). *Comput. Methods Appl. Mech. Engrg.*, **195**, (2006), 1608–1620.

[40] G. Sangalli, Robust a-posteriori estimator for advection-diffusion-reaction problems. *Math. Comp.*, **77**, (2008), 41–70.

[41] D. L. Scharfetter and H. K. Gummel, Large-signal analysis of a silicon read diode oscillator. *IEEE Trans. Electron Devices*, **16**, (1969), 64–77.

[42] J. Schöberl, A posteriori error estimates for Maxwell equations. *Math. Comp.*, **77**, (2008), 633–649.

[43] L. L. Schumaker, *Spline Functions: Basic Theory.* Cambridge University Press, third edition, 2007.

[44] A. Weil, Sur les théorèmes de de Rham. *Comment. Math. Helv.*, **26**, (1952), 119–145.

[45] H. Whitney, *Geometric Integration Theory.* Princeton University Press, Princeton, N. J., 1957.

[46] J. Xu and L. Zikatanov, A monotone finite element scheme for convection-diffusion equations. *Math. Comp.*, **68**, (1999), 1429–1446.

4

On the Complexity of Computing Quadrature Formulas for SDEs

Steffen Dereich
Fachbereich Mathematik und Informatik
Westfälische Wilhelms-Universität Münster

Thomas Müller-Gronbach
Fakultät für Informatik und Mathematik
Universität Passau

Klaus Ritter
Fachbereich Mathematik
Technische Universität Kaiserslautern

Abstract

We survey recent results on the constructive approximation of the distribution of the solution of a stochastic differential equation (SDE) by probability measures with finite support, i.e., by quadrature formulas with positive weights summing up to one. Here we either consider the distribution on the path space or a marginal distribution on the state space. We provide asymptotic results on the N-th minimal error of deterministic and randomized algorithms, which is the smallest error that can be achieved by any such algorithm not exceeding the cost bound N.

4.1 Introduction

Consider an SDE

$$dX(t) = a(X(t))\,dt + b(X(t))\,dW(t), \qquad t \in [0,1],$$

driven by a Brownian motion W and, for simplicity, with a deterministic initial value $X(0) = x_0$. Assuming at least existence of a weak solution and uniqueness in distribution, we study the constructive approximation of

- the distribution $\mu = P_X$ of the solution on the path space $\mathfrak{X}$ or

- the marginal distribution $\mu = P_{X(1)}$ of the solution at time $t = 1$ on the state space $\mathfrak{X}$

by probability measures $\hat{\mu}$ with finite support. Hence

$$\hat{\mu} = \sum_{i=1}^{n} c_i \cdot \delta_{x_i}$$

with $n \in \mathbb{N}$, $x_1, \ldots, x_n \in \mathfrak{X}$ and $c_1, \ldots, c_n > 0$ such that $\sum_{i=1}^{n} c_i = 1$.

In both cases μ is given only implicitly, which constitutes a major challenge. Obviously $\hat{\mu}$ yields a quadrature formula

$$\int_{\mathfrak{X}} f \, d\hat{\mu} = \sum_{i=1}^{n} c_i \cdot f(x_i)$$

for the integral $\int_{\mathfrak{X}} f \, d\mu$ of any μ-integrable function $f\colon \mathfrak{X} \to \mathbb{R}$.

We study deterministic as well as randomized algorithms in a worst case analysis with respect to classes of SDEs, i.e., classes of parameters (x_0, a, b). Typically, these classes are defined in terms of smoothness and degeneracy constraints for the drift coefficient a and the diffusion coefficient b, and x_0 is assumed to belong to a bounded set in $\mathbb{R}^d$. The coefficients a and b are accessible only via an oracle that provides function values or derivative values of these functions at arbitrary points in $\mathbb{R}^d$, and the real number model is used to define the notion of an algorithm.

The worst case error $e(\hat{S})$ of an algorithm $\hat{S}$ is defined in terms of a metric on the space of probability measures on $\mathfrak{X}$, and we are interested in algorithms $\hat{S}$ with an (almost) optimal relation between $e(\hat{S})$ and $\mathrm{cost}(\hat{S})$, the worst case computational cost. To this end we study the N-th minimal errors

$$e_N^{\mathrm{ran}} = \inf\{e(\hat{S}) : \hat{S} \text{ randomized algorithm with } \mathrm{cost}(\hat{S}) \leq N\}$$

and

$$e_N^{\mathrm{det}} = \inf\{e(\hat{S}) : \hat{S} \text{ deterministic algorithm with } \mathrm{cost}(\hat{S}) \leq N\}.$$

Actually we survey asymptotic upper and lower bounds for these quantities as well as algorithms that achieve the upper bounds. The results and the techniques are very different in the cases $\dim(\mathfrak{X}) = \infty$ and $\dim(\mathfrak{X}) < \infty$, i.e., whether we aim at the distribution on the path space or at a marginal distribution on the state space.

The problem of approximating a probability measure on $\mathfrak{X}$ by a probability measure with finite support with respect to a Wasserstein metric is called the quantization problem, see, e.g., [7, 10, 28]. We use this terminology for a general metric as well, and accordingly construction of quadrature formulas may also be called constructive quantization.

We briefly outline the content of this survey. In Section 4.2 we introduce the basic notions of algorithms, error and cost for the construction of quadrature formulas. In Section 4.3 we study deterministic algorithms for approximation on the path space, see [5, 6, 18, 19, 24, 29], while Section 4.4 is devoted to the analogous problem on the state space, see [3, 14, 15, 16, 21, 25]. In Section 4.5 we analyze randomized algorithms for marginal distributions on the state space by means of a randomized quantization technique, see [9].

In the sequel we use $a_n \approx b_n$ to denote the strong asymptotic equivalence of sequences of positive reals, i.e., $\lim_{n\to\infty} a_n/b_n = 1$. Furthermore, $a_n \preceq b_n$ means $a_n \leq \kappa\, b_n$ with a constant $\kappa > 0$. Finally $a_n \asymp b_n$ is used to denote the weak asymptotic equivalence, i.e., $a_n \preceq b_n$ and $b_n \preceq a_n$.

4.2 Algorithms, error, and cost

We informally introduce the notions of deterministic and randomized algorithms, and we define their cost and their error for approximation of P_X or $P_{X(1)}$ by measures with finite support. Actually, we are interested in the approximation problem for a whole class of SDEs, which is defined by constraints on the parameters x_0, a and b. We assume that

$$(x_0, a, b) \in H = H_0 \times H_1 \times H_2,$$

where $H_0 \subset \mathbb{R}^d$ and where H_1 and H_2 are classes of functions $\mathbb{R}^d \to \mathbb{R}^d$ and $\mathbb{R}^d \to \mathbb{R}^{d\times d}$, resp., that are at least Lipschitz continuous. Hence $\mathrm{E}\,\|X\|_\infty^s < \infty$ for every SDE with $(x_0, a, b) \in H$ and every $s \geq 1$.

We take $\mathfrak{X} = C([0,1], \mathbb{R}^d)$ or $\mathfrak{X} = L_p([0,1], \mathbb{R}^d)$ for $1 \leq p < \infty$ with the corresponding norms, if we study approximation of the distribution on the path space, and we take $\mathfrak{X} = \mathbb{R}^d$, if we study approximation of the marginal distribution at $t = 1$. Finally, we use $M(\mathfrak{X})$ to denote the space of Borel probability measures μ on $\mathfrak{X}$ such that

$$\int_{\mathfrak{X}} \|x\|_{\mathfrak{X}}^s \, d\mu(x) < \infty$$

for every $s \geq 1$, and we consider a metric ρ on this space. We study the mapping

$$S\colon H \to M(\mathfrak{X}),$$

where

$$S(x_0, a, b) = P_X \qquad \text{or} \qquad S(x_0, a, b) = P_{X(1)}.$$

In the framework of information-based complexity S is the so-called solution operator, see [33]. Our computational task is to approximate S by means of algorithms that yield measures in

$$M_0(\mathfrak{X}) = \{\mu \in M(\mathfrak{X}) : |\operatorname{supp}(\mu)| < \infty\}.$$

In particular, $(x_0, a, b) \in H$ is the input to any algorithm, but a and b are accessible only via an oracle that provides function values or derivative values of these coefficients at arbitrary points in $\mathbb{R}^d$. The output of an algorithm, assuming termination, is a list $c_1, x_1, \dots, c_n, x_n$, where $c_1, \dots, c_n$ are probability weights and $x_1, \dots, x_n$ are points in $\mathfrak{X}$, suitably coded if $\dim \mathfrak{X} = \infty$. Computations with real numbers and with elementary functions like exp and floor are carried out exactly, and an algorithm has access to a perfect random number generator for the uniform distribution on $[0, 1]$. For a formal definition of this model of computation we refer to [26].

Accordingly, to any algorithm we associate a mapping

$$\hat{S}\colon H \times \Omega \to M_0(\mathfrak{X})$$

given by

$$\hat{S}((x_0, a, b), \omega) = \sum_{i=1}^{n} c_i \cdot \delta_{x_i}.$$

Here $(\Omega, \mathfrak{A}, Q)$ is an underlying computational probability space that carries an independent sequence of random variables, each uniformly distributed on $[0, 1]$, which is used to model the random number generator. Moreover, n as well as the c_i and x_i are random quantities, which depend on (x_0, a, b), too. We put

$$\hat{S}(x_0, a, b) = \hat{S}((x_0, a, b), \cdot)\colon \Omega \to M_0(\mathfrak{X}),$$

and, in an abuse of notation, we also use $\hat{S}$ to denote the randomized algorithm itself.

The cost of applying $\hat{S}$ to (x_0, a, b) is a random quantity, which is denoted by $\operatorname{cost}(\hat{S}, (x_0, a, b))$. It is given by the sum of the number of evaluations of a, b, a', b' etc., the number of arithmetical operations etc. and the number of calls of the random number generator that are carried out by $\hat{S}$ for this input. In particular, we have

$$\operatorname{cost}(\hat{S}, (x_0, a, b)) \geq 2 \cdot |\operatorname{supp}(\hat{S}(x_0, a, b))| \tag{4.1}$$

for all $(x_0, a, b) \in H$. The worst case cost of $\hat{S}$ on H is defined by

$$\operatorname{cost}(\hat{S}) = \sup_{(x_0,a,b)\in H} \mathrm{E}(\operatorname{cost}(\hat{S}, (x_0, a, b))),$$

assuming measurability. Likewise the error of applying $\hat{S}$ to (x_0, a, b) is a random quantity, and the worst case error of $\hat{S}$ on H with respect to the metric ρ is defined by

$$e(\hat{S}) = \sup_{(x_0,a,b)\in H} \left(\mathrm{E}(\rho^2(S(x_0, a, b), \hat{S}(x_0, a, b)))\right)^{1/2},$$

assuming measurability.

Deterministic algorithms are a particular case of randomized algorithms, namely, $\hat{S}$ is not calling the random number generator for any $(x_0, a, b) \in H$, which implies that $\hat{S}(x_0, a, b)$ is a constant mapping for every (x_0, a, b). In the definition of $\operatorname{cost}(\hat{S})$ and $e(\hat{S})$ the expectation may therefore be dropped.

The key quantities are the N-th minimal errors

$$\begin{aligned} e_N^{\mathrm{ran}} &= e_N^{\mathrm{ran}}(H, \mathfrak{X}, \rho) \\ &= \inf\{e(\hat{S}) : \hat{S} \text{ randomized algorithm with } \operatorname{cost}(\hat{S}) \le N\} \end{aligned}$$

of randomized algorithms and the N-th minimal errors

$$\begin{aligned} e_N^{\mathrm{det}} &= e_N^{\mathrm{det}}(H, \mathfrak{X}, \rho) \\ &= \inf\{e(\hat{S}) : \hat{S} \text{ deterministic algorithm with } \operatorname{cost}(\hat{S}) \le N\} \end{aligned}$$

of deterministic algorithms. Clearly $e_N^{\mathrm{ran}} \le e_N^{\mathrm{det}}$.

Concerning the metric ρ on $M(\mathfrak{X})$ the following two cases will be considered in the sequel. Let $\mu_1, \mu_2 \in M(\mathfrak{X})$. The Wasserstein metric $\rho^{(s)}$ of order $1 \le s < \infty$ is defined by

$$\rho^{(s)}(\mu_1, \mu_2) = \inf_{\xi} \left(\int_{\mathfrak{X}\times\mathfrak{X}} \|x_1 - x_2\|_{\mathfrak{X}}^s \, d\xi(x_1, x_2) \right)^{1/s},$$

where the infimum is taken over all Borel probability measures ξ on $\mathfrak{X} \times \mathfrak{X}$ with marginals μ_1 and μ_2, respectively. Furthermore, we study metrics

$$\rho_F(\mu_1, \mu_2) = \sup_{f\in F} \left| \int_{\mathfrak{X}} f \, d\mu_1 - \int_{\mathfrak{X}} f \, d\mu_2 \right|$$

given in a dual representation in terms of a class F of functions $f \colon \mathfrak{X} \to \mathbb{R}$ that satisfies

$$\sup_{f\in F} \sup_{x\in\mathfrak{X}} \frac{|f(x) - f(0)|}{1 + \|x\|_{\mathfrak{X}}^s} < \infty$$

for some $1 \le s < \infty$. In particular, ρ_F is called a metric with ζ-structure, if all functions $f \in F$ are bounded, see [31, p. 72].

Consider $F = \text{Lip}(1)$, i.e., F consists of all functions $f\colon \mathfrak{X} \to \mathbb{R}$ with

$$|f(x_1) - f(x_2)| \le \|x_1 - x_2\|_{\mathfrak{X}}, \qquad x_1, x_2 \in \mathfrak{X}.$$

Then, by the Kantorovich-Rubinstein Theorem,

$$\rho_{\text{Lip}(1)} = \rho^{(1)}.$$

Remark 4.1 We relate our computational problem to an approximation theoretical question. For $\mu \in M(\mathfrak{X})$ and $n \in \mathbb{N}$ let

$$q_n(\mu) = q_n(\mu, \rho) = \inf\{\rho(\mu, \hat{\mu}) : \hat{\mu} \in M_0(\mathfrak{X}),\ |\operatorname{supp}(\hat{\mu})| \le n\}.$$

If $\rho = \rho^{(s)}$ then $q_n(\mu)$ is called the n-th quantization number of μ, see [7, 10, 28]. We use this terminology for a general metric ρ, and accordingly construction of quadrature formulas may also be called constructive quantization.

Consider a randomized algorithm $\hat{S}$ with $\text{cost}(\hat{S}) \le N$, let $(x_0, a, b) \in H$, and put $A = \{\text{cost}(\hat{S}, (x_0, a, b)) \le 2N\}$. Clearly, $Q(A) \ge 1/2$ and $|\operatorname{supp}(\hat{S}(x_0, a, b))| \le N$ on A due to (4.1). Therefore

$$e_N^{\text{ran}} \ge \tfrac{1}{2} \cdot \sup_{(x_0,a,b)\in H} q_N(S(x_0, a, b)). \tag{4.2}$$

Since the asymptotic behavior of the quantization numbers is known in many cases, we may obtain lower bounds for the minimal errors in this way. Note that this argument does not at all exploit the fact that algorithms only have partial information about a and b in the following sense. For every algorithm $\hat{S}$ the number of evaluations of a or b is bounded from above by $\text{cost}(\hat{S}, (x_0, a, b))$, and for a non-trivial class H a finite number of such evaluations does not suffice to determine $S(x_0, a, b)$ exactly. We refer to [30] for lower bounds based on partial information about the drift coefficient a.

Remark 4.2 Consider the marginal case $\mathfrak{X} = \mathbb{R}^d$ and let $\rho = \rho_F$. We compare the construction of quadrature formulas and the actual quadrature problem on F. For the latter task also an oracle for $f \in F$ is needed, and algorithms yield real numbers as output. The worst case error of a randomized algorithm $\tilde{S}\colon H \times F \times \Omega \to \mathbb{R}$ is defined by

$$\Delta(\tilde{S}) = \sup_{(x_0,a,b)\in H, f\in F} \left(\mathrm{E}\Big(\int_{\mathfrak{X}} f \, dS(x_0, a, b) - \tilde{S}(x_0, a, b, f) \Big)^2 \right)^{1/2},$$

and the number of evaluations of f has to be added as another term in the definition of the computational cost. The minimal errors Δ_N^{ran} and Δ_N^{det} for the quadrature problem are defined in the canonical way.

Clearly, every algorithm $\hat{S}$ for construction of quadrature formulas yields an algorithm $\tilde{S}$ for quadrature by

$$\tilde{S}((x_0, a, b), f, \omega) = \int_{\mathfrak{X}} f \, d\hat{S}((x_0, a, b), \omega),$$

and we have $\Delta(\tilde{S}) \leq e(\hat{S})$ as well as $\mathrm{cost}(\hat{S}) \leq \mathrm{cost}(\tilde{S}) \leq 3 \,\mathrm{cost}(\hat{S})$. Hence

$$\Delta_{3N}^{\mathrm{ran}} \leq e_N^{\mathrm{ran}} \qquad \text{and} \qquad \Delta_{3N}^{\mathrm{det}} \leq e_N^{\mathrm{det}}.$$

We stress that the minimal errors $\Delta_{3N}^{\mathrm{ran}}$ and e_N^{ran} may differ substantially, because of the order of taking suprema over F and expectations. For instance, let $H_0 \subset \mathbb{R}^d$ be a bounded set, let

$$H_1 = \{h \in C^1(\mathbb{R}^d, \mathbb{R}^d) \colon |h(0)|, \|\nabla h\|_\infty \leq K\}$$

with $K > 0$, and let H_2 be defined analogously for functions with values in $\mathbb{R}^{d \times d}$. Moreover, let $F = \mathrm{Lip}(1)$. By a well-known result, the Monte Carlo Euler scheme yields

$$\Delta_N^{\mathrm{ran}} \preceq N^{-1/3},$$

while

$$e_N^{\mathrm{ran}} \succeq N^{-1/d}$$

follows from $q_n(\mu) \asymp n^{-1/d}$ for every normal distribution $\mu \in S(H)$ with full support in $\mathbb{R}^d$, see [10, p. 78] and (4.2). We therefore conclude that, at least for $d \geq 4$, the construction of quadrature formulas is substantially harder than quadrature, if randomized algorithms may be used in both cases.

We do not discuss quadrature problems on the path space here, since this requires introducing an appropriate cost model for evaluating functions on infinite dimensional spaces $\mathfrak{X}$, see [2, 13].

Remark 4.3 Precomputing is permitted in our model of computation, and it is actually used in a number of rather different algorithms for the construction of quadrature formulas; we refer to the subsequent sections. Precomputing allows to solve auxiliary problems that do not involve the parameters x_0, a and b of the SDE beforehand. The computational effort for precomputing is not taken into account by $\mathrm{cost}(\hat{S})$.

Remark 4.4 In this paper we discuss constructive approximation by probability measures with finite support. More generally, one might admit signed measures with finite support, which is motivated by, e.g., the stochastic multi-level construction, or one might omit any constraint, which means that $\hat{S}\colon H \times \Omega \to M(\mathfrak{X})$. It seems that these variants of the constructive approximation problem for the distribution of SDEs have not been studied so far.

4.3 Quadrature formulas on the path space

We focus on scalar SDEs, i.e., $d = 1$, and we assume that $H_0 \subset \mathbb{R}$ is a bounded set and

$$H_1 = H_2 = \{h \in C^2(\mathbb{R}) : |h(0)|, \|h'\|_\infty, \|h''\|_\infty \leq K\}$$

with $K > 0$. Moreover, $\mathfrak{X} = \mathfrak{X}_p$ where $\mathfrak{X}_\infty = C([0,1])$ and $\mathfrak{X}_p = L_p([0,1])$ for $1 \leq p < \infty$, and $\rho = \rho^{(s)}$ is the Wasserstein metric of order s on $M(\mathfrak{X}_p)$ for any $1 \leq s < \infty$.

At first we discuss the quantization problem, see Remark 4.1. We assume $b(x_0) \neq 0$ in order to exclude deterministic equations.

Theorem 4.5 *For every $(x_0, a, b) \in H$ with $b(x_0) \neq 0$ there exists a constant $\kappa(x_0, a, b, p, s) > 0$ such that*

$$q_n(S(x_0, a, b)) \approx \kappa(x_0, a, b, p, s) \cdot (\ln n)^{-1/2}.$$

In the particular case of a Brownian motion, i.e., for the parameters $(x_0, a, b) = (0, 0, 1)$, this result is due to [4, 17] for $p = 2$ and to [8] for general p. For SDEs the result is established in [5, 6]. The structure of the asymptotic constant $\kappa(x_0, a, b, p, s)$ and its dependence on the parameters of the SDE is as follows. There exist constants $c(p) > 0$ such that

$$\kappa(x_0, a, b, p, s) = c(p) \cdot \left(\mathrm{E}\left(\int_0^1 |b(X(t))|^{2/(1+2/p)}\, dt \right)^{s(1+2/p)/2} \right)^{1/s},$$

where $1/\infty = 0$. In particular, $c(2) = \sqrt{2}/\pi$, while only upper and lower bounds for $c(p)$ are known in the case $p \neq 2$.

From Theorem 4.5 we immediately get a lower bound for e_N^{ran}, see (4.2), which turns out to be sharp and attainable already by deterministic algorithms.

Theorem 4.6 *We have*

$$e_N^{\text{ran}} \asymp e_N^{\text{det}} \asymp (\ln N)^{-1/2}.$$

The upper bound $e_N^{\text{det}} \preceq (\ln N)^{-1/2}$ in Theorem 4.6 is due to [24], and we describe the deterministic algorithm that yields this bound in the case $p = s = 2$.

Fix $(x_0, a, b) \in H$. The construction is motivated by the following two-level approximation of the solution process X of the corresponding SDE. For $m \in \mathbb{N}$ we consider the Milstein scheme Y with step-size $1/m$ and with piecewise linear interpolation, and we put $t_k = k/m$. Furthermore, we consider the Brownian bridges

$$B_k(t) = W(t) - W(t_k) - (t - t_k) \cdot m \cdot (W(t_{k+1}) - W(t_k)),$$

where $t \in [t_k, t_{k+1}]$. Note that $Y, B_0, \ldots, B_m$ are independent. As an approximation to X we consider $Y + Z$ with the coarse level approximation Y and with a local refinement

$$Z(t) = \sum_{k=0}^{m-1} b(Y(t_k)) \cdot B_k(t) \cdot 1_{[t_k, t_{k+1}]}(t).$$

Instead of the Gaussian measures that are involved in Y and Z the algorithm employs probability measures with finite support.

For the standard normal distribution γ and the Wasserstein metric ρ of order two on $M(\mathbb{R})$ we have $q_n(\gamma, \rho) \asymp n^{-1}$, and we take a sequence of uniform distributions $\hat{\gamma}_r \in M_0(\mathbb{R})$ with $|\operatorname{supp}(\hat{\gamma}_r)| = r$ and

$$\rho(\gamma, \hat{\gamma}_r) \asymp r^{-1}.$$

We refer to [10, Sec. 7.3] for a general account and to [24] for details and additional requirements concerning $\hat{\gamma}_r$ in the present context.

For the distribution ξ of a Brownian bridge on $[0, 1]$ and the Wasserstein metric $\rho^{(2)}$ of order two on $M(L_2([0, 1]))$ we have $q_n(\xi, \rho^{(2)}) \approx \sqrt{2}/\pi \cdot (\ln n)^{-1/2}$, cf. Theorem 4.5, and we take a sequence of measures $\hat{\xi}_n \in M_0(L_2([0, 1]))$ with $|\operatorname{supp}(\hat{\xi}_n)| = n$ and

$$\rho^{(2)}(\xi, \hat{\xi}_n) \asymp (\ln n)^{-1/2}. \tag{4.3}$$

The construction of $\hat{\xi}_n$ is based on the Karhunen-Loève expansion of the Brownian bridge, where the eigenfunctions and eigenvalues are given by $e_\ell(t) = \sqrt{2} \sin(\ell \pi t)$ and $\lambda_\ell = (\ell \pi)^{-2}$, respectively, and on quantization of normal distributions. In particular, $\operatorname{supp}(\hat{\xi}_n) \subset \operatorname{span}\{e_\ell \colon \ell \in \mathbb{N}\}$. In order to achieve

$$\rho^{(2)}(\xi, \hat{\xi}_n) \approx q_n(\xi, \rho^{(2)}) \tag{4.4}$$

suitable quantizations of multi-dimensional centered normal distributions with covariance matrices of diagonal form are used for quantization of blocks of coefficients of the Karhunen-Loève expansion, and this approach involves large scale numerical optimization. Alternatively, (4.3) with reasonably good constants can already be achieved by product quantizers, which merely rely on quantizations of the one-dimensional standard normal distribution. We refer to [7, 19, 20] for surveys and recent results and to the web site [27] for downloads. In any case the construction of suitable measures $\hat{\xi}_n$ is a matter of precomputing.

We first explain how to approximate the distribution P_Y. The standard normal distribution on $\mathbb{R}^m$, which is the basis for the Milstein scheme, is replaced by the uniform distribution

$$\hat{\gamma} = \bigotimes_{k=0}^{m-1} \hat{\gamma}_r.$$

Note that $\operatorname{supp}(\hat{\gamma})$ is a non-uniform grid in $\mathbb{R}^m$. For every $w \in \operatorname{supp}(\hat{\gamma})$ we define a function $y(\cdot; w) \in C([0,1])$ by $y(0; w) = x_0$,

$$\begin{aligned} y(t_{k+1}; w) = y(t_k; w) + 1/m \cdot a(y(t_k; w)) + 1/\sqrt{m} \cdot b(y(t_k; w)) \cdot w_k \\ + 1/(2m) \cdot (b \cdot b')(y(t_k; w)) \cdot (w_k^2 - 1) \end{aligned}$$

and by piecewise linear interpolation. This corresponds to the Milstein scheme with the normalized Brownian increments replaced by the components of w, and as an approximation to the distribution P_Y of the Milstein scheme we use the uniform distribution

$$\hat{\nu} = \frac{1}{r^m} \sum_{w \in \operatorname{supp}(\hat{\gamma})} \delta_{y(\cdot; w)}$$

on a finite set of piecewise linear functions.

Next we turn to the approximation of the distribution P_Z. For every polygon $y \in \operatorname{supp}(\hat{\nu})$ the distribution of the weighted Brownian bridge $b(y(t_k)) \cdot B_k$ on $[t_k, t_{k+1}]$ is approximated by one of the measures $\hat{\xi}_n$, properly rescaled and shifted. Hereby, the size n of the support is chosen in such a way that the local regularity of the solution process X is taken into account. Essentially, we take

$$n = n_k(y) = \max\big(\lfloor M^{\eta_k(y)} \rfloor, 1\big)$$

with

$$\eta_k(y) = |b(y(t_k))| / \sum_{j=0}^{m-1} |b(y(t_j))|$$

for a given $M \in \mathbb{N}$, see [24] for technical details. This strategy, which is crucial for the overall performance of the algorithm, is similar to asymptotically optimal step-size control for pathwise approximation of SDEs, see [11, 12, 22, 23]. To approximate the joint distribution of the weighted Brownian bridges $b(y(t_k)) \cdot B_k$ we take the product measure

$$\hat{\xi}(\cdot, y) = \bigotimes_{k=0}^{m-1} \hat{\xi}_{n_k(y)}(\cdot)$$

in $C([0,1])^m$ and use the linear mapping $\psi(\cdot; y)\colon \operatorname{supp}(\hat{\xi}(\cdot, y)) \to C([0,1])$ given by

$$\psi(z; y)(t) = \sum_{k=0}^{m-1} b(y(t_k)) \cdot z_k((t - t_k) \cdot m)/\sqrt{m} \cdot 1_{[t_k, t_{k+1}]}(t).$$

Choose $m = r \asymp (\ln M)^\alpha$ for $\alpha \in \,]1/2, 1[$, and put $N = m^m \cdot M$. As an approximation to P_{Y+Z}, and thus to P_X, we take

$$\hat{S}_N(x_0, a, b) = \sum_{y \in \operatorname{supp}(\hat{\nu})} \; \sum_{z \in \operatorname{supp}(\hat{\xi}(\cdot, y))} \frac{\hat{\xi}(\{z\}, y)}{r^m} \cdot \delta_{y + \psi(z; y)}.$$

Clearly

$$|\operatorname{supp}(\hat{S}_N(x_0, a, b))| \leq N,$$

and $\hat{\xi}(\{z\}, y)$ is the product of certain probability weights of the measures $\hat{\xi}_n$. The piecewise linear functions y and the functions $\psi(z; y)$, which consist of trigonometric polynomials on the subintervals $[t_k, t_{k+1}]$ can be coded in a natural way. It turns out that $\operatorname{cost}(\hat{S}_N) \preceq m \cdot N$ and $e(\hat{S}_N) \asymp (\ln N)^{-1/2} \asymp (\ln(\operatorname{cost}(\hat{S}_N))^{-1/2}$. Moreover, if (4.4) is satisfied then we even have

$$\rho^{(2)}(S(x_0, a, b), \hat{S}_N(x_0, a, b)) \approx \kappa(x_0, a, b, 2, 2) \cdot (\ln N)^{-1/2} \tag{4.5}$$

for every $(x_0, a, b) \in H$ with $b(x_0) \neq 0$, i.e., the algorithm $\hat{S}_N$ achieves strong asymptotic optimality for the quantization problem for every such SDE, see Remark 4.1 and Theorem 4.5. We refer to [32] for an efficient implementation of $\hat{S}_N$ and to [24] for the construction of $\hat{S}_N$ in the case $p \neq 2$ or $s \neq 2$.

Remark 4.7 The algorithm $\hat{S}_N$ can be generalized to handle systems of SDEs, i.e., to the case $d > 1$, in a straightforward manner. For coarse level approximation, however, it no longer suffices to approximate Brownian increments, since also multiple Itô integrals are needed in the

Milstein scheme. For this task there are several methods available, e.g., the quantization of an approximation to these integrals based on a suitably truncated Karhunen-Loève expansion, see [36], and an empirical measure approach, see Section 4.5, but the analysis of error and cost for construction of quadrature formulas on the path space seems to be open for $d > 1$.

Remark 4.8 The constructive quantization of diffusion processes was initiated by [18], where suitable quantizers for the driving Brownian motion are used as a building block, which is similar to our approach. The approaches differ, however, with respect to the numerical treatment of the SDE. A key assumption in [18] is strict positivity of the diffusion coefficient, which permits the use of the Lamperti transform. Along this way one has to solve n deterministic ODEs, in general, to get an approximation $\hat{\mu}_n$ with $\text{supp}(\hat{\mu}_n) = n$ and $\rho^{(2)}(P_X, \hat{\mu}_n) \asymp (\ln n)^{-1/2}$. This work is extended to systems of SDEs in [29], where rough path theory is used to establish convergence rates in p-variation and in the Hölder metric for ODE-based quantizations. A different approach is developed in [19], where the mean regularity of stochastic processes is exploited. The construction is based on the expansion of X in terms of the Haar basis, and on the availability of optimal quantizations of the corresponding coefficients. It seems that none of these alternative approaches achieves (4.5) and that the computational cost has not been analyzed so far. As far as we understand, these alternative methods do not achieve a cost bound close to the size of the quantization.

4.4 Quadrature formulas on the state space – deterministic algorithms

We study scalar SDEs, i.e., $d = 1$, and we assume that $H_0 \subset \mathbb{R}$ is a bounded set and

$$H_1 = \{h \in C^4(\mathbb{R})\colon |h(0)|, \|h^{(j)}\|_\infty \leq K \text{ for } j = 1, \dots, 4\},$$
$$H_2 = \left\{h \in H_1\colon \inf_{x \in \mathbb{R}} |h(x)| \geq \varepsilon\right\}$$

with $K > 0$ and $\varepsilon \geq 0$. Thus $H_1 = H_2$ if $\varepsilon = 0$, while $\varepsilon > 0$ corresponds to an additional non-degeneracy constraint on the diffusion coefficient b. Furthermore, $\mathfrak{X} = \mathbb{R}$, and we consider the metric $\rho = \rho_F$, where

$$F = \{f \in C^4(\mathbb{R})\colon |f^{(j)}(x)| \leq K \cdot (1 + |x|^\beta) \text{ for } j = 1, \dots, 4\}$$

with $\beta \geq 0$.

In this setting we have the following upper bounds for the minimal errors of deterministic algorithms, see [25].

Theorem 4.9 *For every $\delta > 0$*

$$e_N^{\det} \preceq \begin{cases} N^{-2/3+\delta} & \text{if } \varepsilon > 0, \\ N^{-1/2+\delta} & \text{if } \varepsilon = 0. \end{cases}$$

The central idea to obtain these upper bounds is as follows. We approximate an Euler scheme with step-size $1/m$ by a Markov chain that stays on a grid of small size and has a sparse transition matrix in order to prevent an exponential explosion of the cost. The transition probabilities are chosen in such a way that the central moments of a single Euler step are close to the corresponding moments of a step of the chain. The approximation to the distribution $P_{X(1)}$ is then obtained as the distribution of the Markov chain after m steps.

Here we present this construction in the case $\varepsilon > 1$. Fix $(x_0, a, b) \in H$, let $\delta > 0$ and $m \in \mathbb{N}$, and define the state space of the Markov chain by $\mathcal{Z} = G \cup \{x_0\}$, where

$$G = \{i \cdot m^{-1/2} : i = -\lceil m^{1/2+\delta} \rceil, \dots, \lceil m^{1/2+\delta} \rceil\}.$$

In order to define the transition probabilities $q_{y,z} = q_{y,z}(x_0, a, b)$ for $y, z \in \mathcal{Z}$, we consider an Euler step of length $1/m$ starting in $y \in \mathcal{Z}$, i.e.,

$$Y^y = y + a(y) \cdot m^{-1} + b(y) \cdot m^{-1/2} \cdot V$$

with a standard normally distributed random variable V. Let

$$z_y = y + a(y) \cdot m^{-1}, \qquad \sigma_y = |b(y)| \cdot m^{-1/2}$$

denote the expected value and the standard deviation of Y^y, and put

$$\bar{z}_y = \lceil z_y \cdot m^{1/2} \rceil \cdot m^{-1/2}, \qquad \bar{\sigma}_y = \lceil \sigma_y \cdot m^{1/2} \rceil \cdot m^{-1/2},$$

which will serve as projections of z_y and σ_y onto G, respectively. Essentially, we replace the Euler step by a step from y to at most six possible positions on G, namely

$$\bar{z}_y, \qquad \bar{z}_y \pm \bar{\sigma}_y, \qquad \bar{z}_y - m^{-1/2}, \qquad \bar{z}_y - m^{-1/2} \pm \bar{\sigma}_y. \tag{4.6}$$

To be more precise, we distinguish the two cases given by

$$\mathcal{Z}_1 = \{y \in \mathcal{Z} : \bar{z}_y - m^{-1/2} - \bar{\sigma}_y, \bar{z}_y + \bar{\sigma}_y \in G\}, \quad \mathcal{Z}_2 = \mathcal{Z} \setminus \mathcal{Z}_1.$$

The points $y \in \mathcal{Z}_2$, where z_y is close to the extremal points of G, are absorbing states, i.e.,

$$q_{y,z} = \begin{cases} 1 & \text{if } z = y, \\ 0 & \text{otherwise.} \end{cases}$$

For $y \in \mathcal{Z}_1$ the points given by (4.6) are members of G, and due to $|b(y)| > 1$ all of them are different. Put

$$u_y = m^{1/2} \cdot (\bar{z}_y - z_y)$$

as well as

$$\vartheta_y^{(1)} = \frac{\sigma_y^2}{2\bar{\sigma}_y^2} + \frac{u_y^2 - 2u_y}{6\bar{\sigma}_y^2 \cdot m}, \qquad \vartheta_y^{(2)} = \frac{\sigma_y^2}{2\bar{\sigma}_y^2} + \frac{u_y^2 - 1}{6\bar{\sigma}_y^2 \cdot m}.$$

Clearly, $0 \le u_y < 1$ and $\vartheta_y^{(j)} \le 1/2$. Moreover, $|b(y)| > 1$ yields $\vartheta_y^{(j)} > 0$. We define

$$q_{y,z} = \begin{cases} (1 - u_y) \cdot (1 - 2\vartheta_y^{(1)}) & \text{if } z = \bar{z}_y, \\ (1 - u_y) \cdot \vartheta_y^{(1)} & \text{if } z = \bar{z}_y \pm \bar{\sigma}_y, \\ u_y \cdot (1 - 2\vartheta_y^{(2)}) & \text{if } z = \bar{z}_y - m^{-1/2}, \\ u_y \cdot \vartheta_y^{(2)} & \text{if } z = \bar{z}_y - m^{-1/2} \pm \bar{\sigma}_y, \\ 0 & \text{otherwise.} \end{cases}$$

Let $Q = (q_{y,z})_{y,z \in \mathcal{Z}}$ denote the resulting transition matrix. We approximate $S(x_0, a, b)$ by the discrete distribution

$$\hat{S}_m(x_0, a, b) = \sum_{z \in \mathcal{Z}} (e \cdot Q^m)_z \cdot \delta_z,$$

where $e = (e_z)_{z \in \mathcal{Z}}$ is given by $e_{x_0} = 1$ and $e_z = 0$ for $z \neq x_0$.

Consider a random variable $\tilde{Y}^y$ that models a single step of the Markov chain with transition matrix Q and starting at $y \in \mathcal{Z}$, i.e., $P(\tilde{Y}^y = z) = q_{y,z}$ for every $z \in \mathcal{Z}$. We then have

$$|E(Y^y - z_y)^p - E(\tilde{Y}^y - z_y)^p| \le c \cdot (1 + |y|^{p+1/\delta}) \cdot m^{-2}$$

for every $p \in \mathbb{N}$, where the constant $c > 0$ only depends on K, δ and p. This estimate is a key property to obtain $e(\hat{S}_m) \preceq m^{-1}$. Moreover, we have $\mathrm{cost}(\hat{S}_m) \asymp m^{3/2+\delta}$, which results from the number of arithmetical operations that are used to compute the matrix-vector product $e \cdot Q^m$.

Remark 4.10 We consider, more generally, the case of d-dimensional systems of SDEs with coefficients that satisfy a smoothness constraint of degree $r \in \mathbb{N}$. Fix $K > 0$. For $\beta \geq 0$ we use F^β to denote the class of functions $f\colon \mathbb{R}^d \to \mathbb{R}$ that have continuous partial derivatives $f^{(\alpha)}$ with

$$|f^{(\alpha)}(x)| \leq K \cdot (1 + |x|^\beta)$$

for every $x \in \mathbb{R}^d$ and every $\alpha \in \mathbb{N}_0^d$ with $1 \leq \sum_{i=1}^d \alpha_i \leq r$. We take $\mathfrak{X} = \mathbb{R}^d$, and we consider the metric ρ_F with $F = F^\beta$.

Assume that $H_0 \subset \mathbb{R}^d$ is bounded,

$$H_1 = \{h \in C^r(\mathbb{R}^d, \mathbb{R}^d)\colon |h(0)| \leq K,\, h_1, \dots, h_d \in F^0\}$$

and H_2 is defined analogously to H_1 for functions $h\colon \mathbb{R}^d \to \mathbb{R}^{d \times d}$ with the additional uniform ellipticity constraint that $|y'h(x)| \geq \varepsilon \cdot |y|$ for all $x, y \in \mathbb{R}^d$.

Assume $r \geq 4$. We conjecture that a construction, similar to the Markov chain for $d = 1$, leads to upper bounds

$$e_N^{\mathrm{det}} \preceq \begin{cases} N^{-(r-2)/(d+2)+\delta} & \text{if } \varepsilon > 0, \\ N^{-(r-2)/(2d+2)+\delta} & \text{if } \varepsilon = 0 \end{cases} \tag{4.7}$$

for the minimal errors of deterministic algorithms for every $\delta > 0$. Note that (4.7) holds true for $d = 1$ and $r = 4$ due to Theorem 4.9.

Alternatively, the connection between SDEs and initial value problems for the associated parabolic PDEs could be employed, and numerical methods for PDEs could be used to construct deterministic quadrature formulas on the state space. It would be interesting to investigate the potential of finite difference methods for the latter problem.

We turn to lower bounds for the minimal errors via quantization numbers, which also hold for randomized algorithms. Consider a probability measure $\mu \in M(\mathbb{R}^d)$ with a continuous Lebesgue density that satisfies suitable decay properties. Then

$$q_n(\mu, \rho_F) \asymp n^{-r/d} \tag{4.8}$$

follows from results on weighted approximation and integration in [34, 35]. In particular, (4.8) holds for any non-degenerate d-dimensional normal distribution $\mu \in S(H)$, which implies

$$e_N^{\mathrm{ran}} \succeq N^{-r/d} \tag{4.9}$$

due to (4.2).

Specifically, for $r = 4$ and $d = 1$ we get a lower bound of order 4 in (4.9), which differs substantially from the upper bounds of order 2/3 or 1/2 in Theorem 4.9. It would be interesting to know whether the fact that any algorithm only has partial information about a and b can be used to improve the lower bound, cf. Remark 4.1.

Remark 4.11 Quadrature formulas on the Wiener space, which are based on paths of bounded variation and are exact for iterated integrals up to a fixed degree m, are introduced in [14, 15] and further developed in [3, 21]. The construction of such formulas is a matter of precomputing. For d-dimensional systems of SDEs with smooth coefficients a and b an approximation to the marginal distribution $P_{X(1)}$ of the solution X is obtained by iteratively solving a collection of ODEs on k non-equidistant time intervals. The error is defined in terms of metrics ρ_F with different spaces F. For $F = \mathrm{Lip}(1)$ an error bound of order $k^{-(m-1)/2}$ is achieved. However, the number of ODEs to be solved grows polynomially in k. To cope with this difficulty a recombination technique has been introduced in [16].

Remark 4.12 Let us compare the results from Section 4.3 (for $s = 1$) and 4.4. In both cases we assume a finite degree of smoothness for the coefficients of the SDE and for the elements in F, together with some boundedness or growth constraints. For the problem on the path space we merely achieve a logarithmic convergence of the minimal errors, and the computational cost of the almost optimal algorithm $\hat{S}_N$ presented in Section 4.3 is only slightly larger than $|\,\mathrm{supp}(\hat{S}_N(x_0, a, b))|$. Furthermore, $\mathrm{supp}(\hat{S}_N(x_0, a, b))$ strongly depends on the SDE, i.e., on x_0, a, and b. For the marginal distribution problem we have a polynomial rate of convergence, and the computational cost of the corresponding algorithm $\hat{S}_m$ is substantially larger than $|\,\mathrm{supp}(\hat{S}_m(x_0, a, b))|$. Furthermore, $\mathrm{supp}(\hat{S}_m(x_0, a, b))$ is essentially a grid that does not depend on the SDE.

4.5 Quadrature formulas on the state space – randomized algorithms

In this section we consider systems of SDEs, and we employ random quantization techniques for the marginal distributions $P_{X(1)}$ to obtain randomized algorithms for the construction of quadrature formulas on the state space.

We first present a general result on quantization by empirical measures. For a probability measure μ on $\mathfrak{X} = \mathbb{R}^d$ and $n \in \mathbb{N}$ we use $\hat{\mu}_n$ to denote the empirical measure that is induced by a sequence of n

independent random variables $Y_1, \dots, Y_n$ with $P_{Y_1} = \mu$, i.e.,

$$\hat{\mu}_n = \frac{1}{n} \sum_{i=1}^{n} \delta_{Y_i}.$$

The following upper bound for the average Wasserstein distance of μ and $\hat{\mu}_n$ is due to [9].

Theorem 4.13 *Let $1 \le s < d/2$ and $q > ds/(d-s)$. There exists a constant $\kappa(d,s,q) > 0$ such that*

$$\left(\mathrm{E}\left(\left(\rho^{(s)}(\mu, \hat{\mu}_n)\right)^s\right)\right)^{1/s} \le \kappa(d,s,q) \cdot \left(\int_{\mathbb{R}^d} |x|^q \, d\mu(x)\right)^{1/q} \cdot n^{-1/d}$$

for every probability measure μ on $\mathbb{R}^d$ and all $n \in \mathbb{N}$.

Note that the bound stated in Theorem 4.13 allows us to control the approximation error in terms of moments of the probability measure μ. Results of this type are commonly referred to as Pierce type estimates in quantization.

We add that the n-th quantization numbers satisfy

$$q_n(\mu, \rho^{(s)}) \asymp n^{-1/d} \tag{4.10}$$

if μ has a finite moment of order $s + \delta$ for some $\delta > 0$ and a nonvanishing absolute continuous part with respect to the Lebesgue measure on $\mathbb{R}^d$. See [10, Thm. 6.2] for further details and for results on the strong asymptotic behavior of $q_n(\mu, \rho^{(s)})$.

In the sequel we consider systems of SDEs with the same constraints as in Remark 4.2. Thus $H_0 \subset \mathbb{R}^d$ is a bounded set,

$$H_1 = \{h \in C^1(\mathbb{R}^d, \mathbb{R}^d) \colon |h(0)|, \|\nabla h\|_\infty \le K\}$$

with $K > 0$, and H_2 is defined analogously for functions with values in $\mathbb{R}^{d \times d}$. Moreover, $\mathfrak{X} = \mathbb{R}^d$, and we consider the Wasserstein metric $\rho^{(s)}$ of order $1 \le s < d/2$ on $M(\mathfrak{X})$.

Theorem 4.14 *For $d \ge 5$ we have*

$$N^{-1/d} \preceq e_N^{\mathrm{ran}} \preceq N^{-1/(d+2)}.$$

The lower bound in this result is a consequence of (4.2) and (4.10), and in fact it holds for every $d \in \mathbb{N}$ and $s \ge 1$.

Note that Theorem 4.13 is not directly applicable to derive an upper bound for the minimal error e_N^{ran}, since one cannot sample from the marginal distribution $P_{X(1)}$ in general. More precisely, there is no randomized algorithm $A : H \times \Omega \to \mathbb{R}^d$, which uses partial information about a and b, such that the distribution of $A((x_0, a, b), \cdot)$ coincides with $P_{X(1)}$ for every $(x_0, a, b) \in H$.

We show that the upper bound in Theorem 4.14 is obtained by the empirical measure based on the Euler scheme. Let $m, n \in \mathbb{N}$, and let $V_1, \ldots, V_m$ denote an independent sequence of standard normally distributed d-dimensional random vectors. Fix $(x_0, a, b) \in H$. Put $Y^{(0)} = x_0$ and

$$Y^{(k+1)} = Y^{(k)} + a(Y^{(k)}) \cdot m^{-1} + b(Y^{(k)}) \cdot m^{-1/2} \cdot Z^{(k)}$$

for $k = 0, \ldots, m-1$, and take n independent copies $Y_1^{(m)}, \ldots, Y_n^{(m)}$ of $Y^{(m)}$. We define

$$\hat{S}_{m,n}((x_0, a, b), \omega) = \frac{1}{n} \sum_{i=1}^{n} \delta_{Y_i^{(m)}(\omega)}.$$

We briefly analyze the error and the cost of $\hat{S}_{m,n}$. Fix $(x_0, a, b) \in H$, and put $\mu^{(m)} = P_{Y^{(m)}}$ and $\hat{\mu}_n^{(m)} = \hat{S}_{m,n}(x_0, a, b)$. Then

$$\begin{aligned} &\left(\mathrm{E}\left(\left(\rho^{(s)}(S(x_0, a, b), \hat{S}_{m,n}(x_0, a, b))\right)^2\right)\right)^{1/2} \\ &\qquad \le \rho^{(s)}(P_{X(1)}, P_{Y^{(m)}}) + \left(\mathrm{E}\left(\left(\rho^{(s)}(\mu^{(m)}, \hat{\mu}_n^{(m)})\right)^2\right)\right)^{1/2}. \end{aligned}$$

Put $s^* = \max(s, 2)$ and apply Theorem 4.13 with any $q > ds^*/(d - s^*)$ to obtain

$$\left(\mathrm{E}\left(\left(\rho^{(s)}(P_{Y^{(m)}}, \hat{\mu}_{m,n})\right)^2\right)\right)^{1/2} \le \kappa_1 \cdot n^{-1/d}$$

for every $m \in \mathbb{N}$, with a constant $\kappa_1 > 0$ that only depends on d, s, x_0, K. Similarly, we have

$$\rho^{(s)}(P_{X(1)}, P_{Y^{(m)}}) \le E(|X(1) - Y^{(m)}|^s)^{1/s} \le \kappa_2 \cdot m^{-1/2}, \tag{4.11}$$

for every $m \in \mathbb{N}$, with a constant $\kappa_2 > 0$ that only depends on d, s, x_0, K. Hence

$$e(\hat{S}_{n,m}) \preceq m^{-1/2} + n^{-1/d}.$$

Clearly,

$$\mathrm{cost}(\hat{S}_{m,n}) \le \kappa \cdot d^2 \cdot n \cdot m$$

with a constant $\kappa > 0$. Choose $m \asymp n^{2/d}$ to obtain the estimate in Theorem 4.14.

We add that the upper bound in Theorem 4.14 is valid for $d \ge 3$, if the worst case error $e(\hat{S})$ of an algorithm $\hat{S}$ is defined in terms of moments of order one rather than order two.

Remark 4.15 Under stronger assumptions on the parameters of the SDE, i.e., for smaller classes H, improved upper bounds should hold for e_N^{ran}.

In [1] the Euler scheme is studied for SDEs with C^∞-coefficients that have bounded partial derivatives of any order and satisfy a nondegeneracy condition. For every bounded measurable function $f : \mathbb{R}^d \to \mathbb{R}$ an upper bound of order $1/m$ is obtained for the so-called weak error $\left| \int_{\mathbb{R}^d} f\, dP_{X(1)} - \int_{\mathbb{R}^d} f\, dP_{Y^{(m)}} \right|$. A careful adaptation of [1] should therefore lead to an upper bound of order m^{-1} in (4.11) and, consequently, to an upper bound of order $1/N^{1/(d+1)}$ in Theorem 4.14 in the case $s = 1$ for an appropriate class H.

An alternative approach is to employ Itô-Taylor schemes of higher order instead of the Euler scheme, which leads to faster convergence in (4.11) under appropriate assumptions on H_1 and H_2. Actually, for each $\delta > 0$ one can determine classes of coefficients such that an upper bound of order $1/N^{1/d-\delta}$ is valid in Theorem 4.14. However, this approach makes use of iterated integrals, and the effort for precomputation may therefore be prohibitively large if δ is small.

Acknowledgement

This work was supported by the Deutsche Forschungsgemeinschaft (DFG) within the Priority Programme 1324. We are grateful to Stefan Heinrich, Henryk Woźniakowski and Larisa Yaroslavtseva for stimulating discussions.

References

[1] V. Bally and D. Talay, The law of the Euler scheme for stochastic differential equations (I): convergence rate of the distribution function. *Prob. Theory and Related Fields*, **104**, 43-60, 1995.

[2] J. Creutzig, S. Dereich, T. Müller-Gronbach and K. Ritter, Infinite-dimensional quadrature and approximation of distributions. *Found. Comput. Math.*, **9**, 391–429, 2009.

[3] D. Crisan and S. Ghazali, On the convergence rates of a general class of

weak approximation of SDEs. in: *Stochastic Differential Equations, Theory and Applications*, P. Baxendale, S. V. Lototsky, eds., 221–249, World Scientific, Singapore, 2007.

[4] S. Dereich, *High resolution coding of stochastic processes and small ball probabilities.* Ph. D. Thesis, Department of Mathematics, 2003, FU Berlin.

[5] S. Dereich, The coding complexity of diffusion processes under supremum norm distortion. *Stochastic Processes Appl.*, **118**, 917–937, 2008.

[6] S. Dereich, The coding complexity of diffusion processes under $L^p[0,1]$-norm distortion. *Stochastic Processes Appl.*, **118**, 938–951, 2008.

[7] S. Dereich, Asymptotic formulae for coding problems and intermediate optimization problems: a review. in: *Trends in Stochastic Analysis*, J. Blath, P. Moerters, M. Scheutzow, eds., 187–232, Cambridge Univ. Press, Cambridge, 2009.

[8] S. Dereich and M. Scheutzow, High-resolution quantization and entropy coding for fractional Brownian motion. *Electron. J. Probab.*, **11**, 700–722, 2006.

[9] S. Dereich, M. Scheutzow and R. Schottstedt, Constructive quantization: approximation by empirical measures. Preprint, arXiv:1108.5346, 2011.

[10] S. Graf and H. Luschgy, *Foundations of Quantization for Probability Distributions.* Lect. Notes in Math. **1730**, Springer-Verlag, Berlin, 2000.

[11] N. Hofmann, T. Müller-Gronbach and K. Ritter, The optimal discretization of stochastic differential equations. *J. Complexity*, **17**, 117–153, 2001.

[12] N. Hofmann, T. Müller-Gronbach and K. Ritter, Linear vs. standard information for scalar stochastic differential equations. *J. Complexity*, **18**, 394–414, 2002.

[13] F. Y. Kuo, I. H. Sloan, G. W. Wasilkowski and H. Woźniakowski, Liberating the dimension. *J. Complexity*, **26**, 422–454, 2010.

[14] S. Kusuoka, Approximation of expectation of diffusion process and mathematical finance. *Adv. Stud. Pure Math.*, **31**, 147–165, 2001.

[15] S. Kusuoka, Approximation of expectation of diffusion processes based on Lie algebra and Malliavin calculus. *Adv. Math. Econ.*, **6**, 69–83, 2004.

[16] C. Litterer and T. Lyons, High order recombination and an application to cubature on the Wiener space. Preprint. http://arxiv.org/pdf/1008.4942v1, 2010.

[17] H. Luschgy and G. Pagès, Sharp asymptotics of the functional quantization problem for Gaussian processes. *Ann. Appl. Prob.*, **32**, 1574–1599, 2004.

[18] H. Luschgy and G. Pagès, Functional quantization of a class of Brownian diffusion: a constructive approach. *Stochastic Processes Appl.*, **116**, 310–336, 2006.

[19] H. Luschgy and G. Pagès, Functional quantization rate and mean regularity of processes with an application to Lévy processes. *Ann. Appl. Prob.*, **18**, 427–469, 2008.

[20] H. Luschgy, G. Pagès and B. Wilbertz, Asymptotically optimal quantization schemes for Gaussian processes. *ESAIM Probab. Stat.*, **14**, 93–116, 2008.

[21] T. Lyons and N. Victoir, Cubature on Wiener space. *Proc. Royal Soc. Lond.*, **460**, 169–198, 2004.

[22] T. Müller-Gronbach, Optimal uniform approximation of systems of stochastic differential equations. *Ann. Appl. Prob.*, **12**, 664–690, 2002.

[23] T. Müller-Gronbach, Optimal pointwise approximation of SDEs based on Brownian motion at discrete points. *Ann. Appl. Prob.*, **14**, 1605-1642, 2004.

[24] T. Müller-Gronbach and K. Ritter, A local refinement strategy for constructive quantization of scalar SDEs. Preprint **72**, DFG SPP 1324, 2010.

[25] T. Müller-Gronbach, K. Ritter and L. Yaroslavtseva, A derandomization of the Euler scheme for scalar stochastic differential equations. To appear in *J. Complexity.*

[26] E. Novak, The real number model in numerical analysis. *J. Complexity,* **11**, 57–73, 1995.

[27] G. Pagès and J. Printems, `http://www.quantize.maths-fi.com/`, 2005.

[28] G. Pagès and J. Printems, Optimal quantization for finance: from random vectors to stochastic processes. in: *Mathematical Modelling and Numerical Methods in Finance, Handbook of Numerical Analysis,* Vol. XV, A. Bensoussan, Q. Zhang, eds., 595–648, North-Holland, Amsterdam, 2008.

[29] G. Pagès and A. Sellami, Convergence of multi-dimensional quantized SDE's. Preprint, 2010.

[30] K. Petras and K. Ritter, On the complexity of parabolic initial value problems with variable drift. *J. Complexity,* **22**, 118–145, 2006.

[31] S. T. Rachev, *Probability Metrics and the Stability of Stochastic Models.* Wiley, Chichester, 1991.

[32] S. Toussaint, Konstruktive Quantisierung skalarer Diffusionsprozesse. Diploma Thesis, Department of Mathematics, 2008, TU Darmstadt.

[33] J. F. Traub, G. W. Wasilkowski and H. Woźniakowski, *Information-Based Complexity.* Academic Press, New York, 1988.

[34] G. W. Wasilkowski and H. Woźniakowski, Complexity of weighted approximation over $\mathbb{R}$. *J. Approx. Theory,* **103**, 223–251, 2000.

[35] G. W. Wasilkowski and H. Woźniakowski, Complexity of weighted approximation over $\mathbb{R}^d$. *J. Complexity,* **17**, 722–740, 2001.

[36] M. Wiktorsson, Joint characteristic function and simultaneous simulation of iterated Itô integrals for multiple independent Brownian motions. *Ann. Appl. Prob.*, **11**, 470–487, 2001.

5

The Quantum Walk of F. Riesz

F. Alberto Grünbaum[a]
Department of Mathematics
University of California, Berkeley

Luis Velázquez[b]
Departamento de Matemática Aplicada
Universidad de Zaragoza

Abstract

We exhibit a way to associate a quantum walk on the non-negative integers to any probability measure on the unit circle. This forces us to consider one step transitions that are not traditionally allowed. We illustrate this in the case of a very interesting measure, originally proposed by F. Riesz for a different purpose.

5.1 Introduction and contents of the paper

The purpose of this note is to consider the probability measure constructed by F. Riesz, [18], back in 1918, and to study a quantum walk naturally associated to it.

The measure on the unit circle that F. Riesz built is formally given by the expression

$$d\mu(z) = \prod_{k=1}^{\infty}(1+\cos(4^k\theta))\frac{d\theta}{2\pi} = \prod_{k=1}^{\infty}(1+(z^{4^k}+z^{-4^k})/2)\frac{dz}{2\pi iz} = \sum_{j=-\infty}^{\infty}\overline{\mu}_j z^j\frac{dz}{2\pi iz}\,. \tag{5.1}$$

[a] Supported in part by the Applied Math. Sciences subprogram of the Office of Energy Research,USDOE, under Contract DE-AC03-76SF00098.

[b] Partly supported by the research project MTM2008-06689-C02-01 from the Ministry of Science and Innovation of Spain and the European Regional Development Fund (ERDF), and by Project E-64 of Diputación General de Aragón (Spain).

Here $z = e^{i\theta}$. If one truncates this infinite product the corresponding measure has a nice density. These approximations converge weakly to the Riesz measure.

The recent paper [6] gives a natural path to associate to a quantum walk on the non-negative integers a probability measure on the unit circle. This construction is also pushed to quantum walks on the integers. The traditional class of coined quantum walks considered in the literature allows for certain one step transitions and this leads to a restricted class of probability measures.

In this paper we take the attitude that for an arbitrary probability measure a slightly more general recipe for these transitions gives rise to a quantum walk. The measure considered by Riesz falls outside of the more restricted class considered so far, and is used here as an interesting example.

There is an obvious danger that having gone beyond the traditional class of quantum walks some of the appealing properties of these walks may no longer hold. From this perspective we consider the example discussed here as a laboratory situation where we will test some of these features.

There is an extra reason for looking at this special example: Riesz's measure is one of the nicest examples of purely singular continuous measures. This means that the unitary operator governing the evolution of the corresponding quantum walk has a pure singular continuous spectrum. Hence, Riesz's quantum walk becomes an ideal candidate to analyze the dynamical consequences of such a kind of elusive spectrum.

We first show how to introduce a quantum walk given a probability measure on the unit circle and then we analyze in more detail the case of Riesz's measure, and give some exploratory results pertaining to the large time behaviour of the "site distribution" for this non-standard walk.

We are grateful to Prof. Reinhard Werner for pointing out that F. Riesz actually started the infinite product (5.1) with $k = 0$. There are two well known references [22, 8] that use the convention used here. Each choice has its own advantages as will be seen in section 5.

For a review of Riesz's construction and its many uses, see [28, 22, 11]. For reviews of quantum walks, see [1, 12, 13].

5.2 Szegő polynomials and CMV matrices

Let μ be a probability measure on the unit circle $\mathbb{T} = \{z \in \mathbb{C} : |z| = 1\}$, and $L^2_\mu(\mathbb{T})$ the Hilbert space of μ-square-integrable functions with inner product

$$(f, g) = \int_{\mathbb{T}} \overline{f(z)}\, g(z)\, d\mu(z).$$

For simplicity we assume that the support of μ contains an infinite number of points.

A very natural operator to consider in our Hilbert space is given by

$$\begin{aligned} U_\mu \colon L^2_\mu(\mathbb{T}) &\to L^2_\mu(\mathbb{T}) \\ f(z) &\longrightarrow z f(z). \end{aligned} \tag{5.2}$$

Since the Laurent polynomials are dense in $L^2_\mu(\mathbb{T})$, a natural basis to obtain a matrix representation of U_μ is given by the Laurent polynomials $\{\chi_j\}_{j=0}^{\infty}$ obtained from the Gram–Schmidt orthonormalization of $\{1, z, z^{-1}, z^2, z^{-2}, \dots\}$ in $L^2_\mu(\mathbb{T})$.

The matrix $\mathcal{C} = (\chi_j, z\chi_k)_{j,k=0}^{\infty}$ of U_μ with respect to $\{\chi_j\}$ has the form

$$\mathcal{C} = \begin{pmatrix} \overline{\alpha}_0 & \rho_0\overline{\alpha}_1 & \rho_0\rho_1 & 0 & 0 & 0 & 0 & \dots \\ \rho_0 & -\alpha_0\overline{\alpha}_1 & -\alpha_0\rho_1 & 0 & 0 & 0 & 0 & \dots \\ 0 & \rho_1\overline{\alpha}_2 & -\alpha_1\overline{\alpha}_2 & \rho_2\overline{\alpha}_3 & \rho_2\rho_3 & 0 & 0 & \dots \\ 0 & \rho_1\rho_2 & -\alpha_1\rho_2 & -\alpha_2\overline{\alpha}_3 & -\alpha_2\rho_3 & 0 & 0 & \dots \\ 0 & 0 & 0 & \rho_3\overline{\alpha}_4 & -\alpha_3\overline{\alpha}_4 & \rho_4\overline{\alpha}_5 & \rho_4\rho_5 & \dots \\ 0 & 0 & 0 & \rho_3\rho_4 & -\alpha_3\rho_4 & -\alpha_4\overline{\alpha}_5 & -\alpha_4\rho_5 & \dots \\ \dots & \dots & \dots & \dots & \dots & \dots & \dots & \dots \end{pmatrix},$$

where $\rho_j = \sqrt{1 - |\alpha_j|^2}$ and $\{\alpha_j\}_{j=0}^{\infty}$ is a sequence of complex numbers such that $|\alpha_j| < 1$. The coefficients α_j are known as the Verblunsky (or Schur, or Szegő, or reflection) parameters of the measure μ, and establish a bijection between the probability measures supported on an infinite set of the unit circle and sequences of points in the open unit disk. The unitary matrices of the form above are called CMV matrices, see [22, 23, 27].

The problem of finding the sequence $\{\alpha_j\}$ for a given measure μ or, more generally, that of relating properties of the measure and the sequence is a central problem, see [22], where a few explicit examples are recorded. Even in cases when the measure is a very natural one, this can be a hard problem. Back in the 1980's one of us formulated a conjecture based on work on the limited angle problem in X-ray tomography. The same conjecture was also made in a slightly different context in work of Delsarte, Janssen and deVries. The conjecture amounts to showing that the Verblunsky parameters of a certain measure are all positive. This was finally established in a real tour-de-force in [14].

One of the results of this paper consists of finding these parameters in the case of F. Riesz's measure. In the process of finding these parameters we will need to invoke some other sequences. Some of these will be subsequences of $\{\alpha_j\}$, and some other ones will only have an auxiliary role. We will propose an ansatz for the Verblunsky parameters of the Riesz measure that have been checked so far for the first 6000 non-null Verblunsky parameters. This is enough for computational purposes concerning the related quantum walk. A proof of our ansatz deserves additional efforts.

The decomposition of a measure $d\mu$ above into an absolutely continuous and a singular part can be further refined by splitting the singular part into point masses and a singular continuous part. The example of Riesz that we will consider later will consist only of this third type of measure, and it is (most likely) the first known example of a measure of this kind, built in terms of a formal Fourier series. For the case of the unit interval there is a construction of such a singular continuous measure in the classical book by F. Riesz and B. Sz-Nagy which is most likely due to Lebesgue. Notice that the method of "Riesz products" introduced in [18] can be used to produce measures such as the one that lives in the Cantor middle-third set. However in this case, as well as in the one due to Lebesgue, one loses the tight connection with Fourier analysis that makes the example of Riesz easier to handle.

A very important role will be played by the Carathéodory function F of the orthogonality measure μ, defined by

$$F(z) = \int_{\mathbb{T}} \frac{t+z}{t-z}\, d\mu(t), \qquad |z| < 1.$$

F is analytic on the open unit disc with McLaurin series

$$F(z) = 1 + 2\sum_{j=1}^{\infty} \overline{\mu}_j z^j, \qquad \mu_j = \int_{\mathbb{T}} z^j d\mu(z),$$

whose coefficients provide the moments μ_j of the measure μ.

Another useful tool in the theory of orthogonal polynomials on the unit circle is the so called Schur function related to μ by means of F through the expression

$$f(z) = z^{-1}(F(z)-1)(F(z)+1)^{-1}, \quad |z| < 1.$$

Since the Schur function, and its Taylor coefficients will play such an important role, see [9], we will settle the issue of names by sticking to the name Verblunsky parameters for those that could be also called by the names of Schur or Szegő.

These functions obtained here by starting from a probability measure on $\mathbb{T}$ can be characterized as the analytic functions on the unit disk $\mathbb{D} = \{z \in \mathbb{C} : |z| < 1\}$ such that $F(0) = 1$, Re $F(z) > 0$ and $|f(z)| < 1$ for $z \in \mathbb{D}$, respectively.

Starting from $f_0 = f$, the Verblunsky parameters $\alpha_k = f_k(0)$ can be obtained through the Schur algorithm that produces a sequence of functions $\{f_k\}$ by means of

$$f_{k+1}(z) = \frac{1}{z}\frac{f_k(z) - \alpha_k}{1 - \overline{\alpha}_k f_k(z)}.$$

By using the reverse recursion

$$f_k(z) = \frac{z f_{k+1}(z) + \alpha_k}{1 + \overline{\alpha}_k z f_{k+1}(z)} = \alpha_k + \frac{\rho_k^2}{\overline{\alpha}_k + \frac{1}{z f_{k+1}(z)}}$$

one can obtain a continued fraction expansion for $f(z)$. This is called a "continued fraction-like" algorithm by Schur, [21], and made into an actual one by H. Wall in [26]. See also [22]. We will illustrate the power of this way of computing these parameters by using it in our example to compute (with computer assistance, in exact arithmetic) enough of them so that we can formulate an ansatz as to the form of these parameters.

5.3 Traditional quantum walks

We consider one-dimensional quantum walks with basic states $|i\rangle \otimes |\uparrow\rangle$ and $|i\rangle \otimes |\downarrow\rangle$, where i runs over the non-negative integers, and with a one step transition mechanism given by a unitary matrix U. This is usually done by considering a coin at each site i, as we will see below.

One considers the following dynamics: a spin up can move to the right and remain up or move to the left and change orientation. A spin down can either go to the right and change orientation or go to the left and remain down.

In other words, only the nearest neighbour transitions such that the final spin (up/down) agrees with the direction of motion (right/left) are allowed. This dynamics bears a resemblance to the effect of a magnetic interaction on quantum system with spin: the spin decides the direction of motion. This rule applies to values of the site variable $i \geq 1$ and needs to be properly modified at $i = 0$ to get a unitary evolution.

Schematically, the allowed one step transitions are

$$|i\rangle \otimes |\uparrow\rangle \longrightarrow \begin{cases} |i+1\rangle \otimes |\uparrow\rangle & \text{with amplitude } c_{11}^i \\ |i-1\rangle \otimes |\downarrow\rangle & \text{with amplitude } c_{21}^i \end{cases}$$

$$|i\rangle \otimes |\downarrow\rangle \longrightarrow \begin{cases} |i+1\rangle \otimes |\uparrow\rangle & \text{with amplitude } c_{12}^i \\ |i-1\rangle \otimes |\downarrow\rangle & \text{with amplitude } c_{22}^i \end{cases}$$

where, in the case $i = 0$, the unitarity requirement forces the identification $|-1\rangle \otimes |\downarrow\rangle \equiv |0\rangle \otimes |\uparrow\rangle$. For each $i = 0, 1, 2, \ldots,$

$$C_i = \begin{pmatrix} c_{11}^i & c_{12}^i \\ c_{21}^i & c_{22}^i \end{pmatrix} \tag{5.3}$$

is an arbitrary unitary matrix which we will call the ith coin.

If we choose to order the basic states of our system as follows

$$|0\rangle \otimes |\uparrow\rangle,\ |0\rangle \otimes |\downarrow\rangle,\ |1\rangle \otimes |\uparrow\rangle,\ |1\rangle \otimes |\downarrow\rangle,\ |2\rangle \otimes |\uparrow\rangle,\ |2\rangle \otimes |\downarrow\rangle,\ \ldots \tag{5.4}$$

then the transition matrix is given below

$$U = \begin{pmatrix} c_{21}^0 & 0 & c_{11}^0 & & & & & & \\ c_{22}^0 & 0 & c_{12}^0 & 0 & & & & & \\ 0 & c_{21}^1 & 0 & 0 & c_{11}^1 & & & & \\ & c_{22}^1 & 0 & 0 & c_{12}^1 & 0 & & & \\ & & 0 & c_{21}^2 & 0 & 0 & c_{11}^2 & & \\ & & & c_{22}^2 & 0 & 0 & c_{12}^2 & 0 & \\ & & & & \ddots & \ddots & \ddots & \ddots & \ddots \end{pmatrix}$$

and we take this as the transition matrix for a traditional quantum walk on the non-negative integers with arbitrary (unitary) coins C_i as in (5.3) for $i = 0, 1, 2, \ldots$.

The reader will notice that the structure of this matrix is not too different from a CMV matrix for which the odd Verblunsky parameters vanish. This feature will guarantee that in the CMV matrix the central 2×2 blocks would vanish identically. The CMV matrix should have real and positive entries in some of the 2×1 matrices that are adjacent to the central 2×2 blocks, and this is not generally true for the unitary matrix given above. In [6] one proves that this can be taken care of by an appropriate conjugation with a diagonal matrix. This associates a measure μ on the unit circle with the above matrix U. Indeed, U

becomes the matrix representation of the operator U_μ, defined in (5.2), with respect to an orthonormal basis of Laurent polynomials X_j differing only by constant phase factors $e^{i\theta_j}$ from the standard ones $\{\chi_j\}$ giving the CMV matrix.

In [6] one considers the case of a constant coin C_i for which the measure μ and the function $F(z)$ are explicitly found. In this case, after the conjugation alluded to above, the Verblunsky parameters are given by

$$a, 0, a, 0, a, 0, a, 0 \ldots$$

for a value of a that depends on the coin, and the function $F(z)$ is, up to a rotation of the variable z, given by the function

$$F(z) = -\frac{z - z^{-1} - 2i\mathrm{Im}\ a}{\sqrt{(z - z^{-1})^2 + 4|a|^2} - 2\mathrm{Re}\ a}.$$

The corresponding Schur function is the even function of z

$$f(z) = \frac{z^2 - 1 + \sqrt{(z^2 - 1)^2 + 4|a|^2 z^2}}{2\bar{a}z^2}.$$

It is easy to see that, in general, the condition $f(-z) = f(z)$ is equivalent to requiring that the odd Verblunsky parameters of μ should vanish. Traditional coined quantum walks are therefore those whose Schur function is an even function of z. In terms of the Carathéodory function the restriction to a traditional quantum walk amounts to $F(-z)F(z) = 1$.

5.4 Quantum walks resulting from an arbitrary probability measure

One of the main points of [6] was to show that the use of the measure μ allows one to associate with each state of our quantum walk a complex valued function in $L^2_\mu(\mathbb{T})$ in such a way that the transition amplitude between any two sates in time n is given by an integral with respect to μ involving the corresponding functions and the quantity z^n. More explicitly we have

$$\langle \tilde{\Psi} | U^n | \Psi \rangle = \int_{\mathbb{T}} z^n \psi(z) \overline{\tilde{\psi}(z)}\, d\mu(z),$$

where $\psi(z) = \sum_j \psi_j X_j(z)$ is the $L^2_\mu(\mathbb{T})$ function associated with state $|\Psi\rangle = \sum_j \psi_j |j\rangle$. Here $|j\rangle$ is the j-th vector of the ordered basis consisting of basic vectors as given in (5.4), i.e. $|j\rangle$ stands for a site and a spin orientation, while $X_j(z)$ are the orthonormal Laurent polynomials related to the transition matrix of the quantum walk. Similarly $\tilde{\psi}(z)$ is the function associated to the state $|\Psi\rangle$.

This construction will now be extended to the case of any transition mechanism that is cooked out of a CMV matrix as above. More explicitly, we allow for the following dynamics

$$|i\rangle \otimes |\uparrow\rangle \longrightarrow \begin{cases} |i+1\rangle \otimes |\uparrow\rangle & \text{with amplitude } \rho_{i+2}\rho_{i+3} \\ |i-1\rangle \otimes |\downarrow\rangle & \text{with amplitude } \rho_{i+1}\overline{\alpha}_{i+2} \\ |i\rangle \otimes |\uparrow\rangle & \text{with amplitude } -\alpha_{i+1}\overline{\alpha}_{i+2} \\ |i\rangle \otimes |\downarrow\rangle & \text{with amplitude } \rho_{i+2}\overline{\alpha}_{i+3} \end{cases}$$

$$|i\rangle \otimes |\downarrow\rangle \longrightarrow \begin{cases} |i+1\rangle \otimes |\uparrow\rangle & \text{with amplitude } -\alpha_{i+2}\rho_{i+3} \\ |i-1\rangle \otimes |\downarrow\rangle & \text{with amplitude } \rho_{i+1}\rho_{i+2} \\ |i\rangle \otimes |\uparrow\rangle & \text{with amplitude } -\alpha_{i+1}\rho_{i+2} \\ |i\rangle \otimes |\downarrow\rangle & \text{with amplitude } -\alpha_{i+2}\overline{\alpha}_{i+3}. \end{cases}$$

The expressions for the amplitudes above are valid for any even i with the convention $|-1\rangle \otimes |\downarrow\rangle \equiv |0\rangle \otimes |\uparrow\rangle$. If i is odd then in every amplitude the index i needs to be replaced by $i-1$.

This generalization of the traditional coined quantum walks consists in adding the possibility of self-transitions for each site. One can, in principle, consider even more general transitions. As long as the evolution is governed by a unitary operator with a cyclic vector there is a CMV matrix lurking around. In our case the basis is given directly in terms of the basic states and there is no need to look for a new basis.

It is clear that the main results in [6] extend to this more general case. If we have a way of computing the orthogonal Laurent polynomials we get an integral expression for the transition amplitude for going between any pair of basic states in any number of steps.

5.5 The Schur function for Riesz's measure

From the expression for the Riesz measure given earlier we see that the expansion

$$d\mu(z) = \sum_{j=-\infty}^{\infty} \overline{\mu}_j z^j \frac{dz}{2\pi i z}$$

leads to the moments μ_j of the measure μ. Apart form the first one, $\mu_0 = 1$, if $j \neq 0$ can be written, in the necessarily unique form, as

$$j = \pm 4^{k_1} \pm 4^{k_2} \pm \cdots \pm 4^{k_p}, \qquad k_1 > k_2 > \cdots > k_p \geq 1,$$

then

$$\mu_j = 1/2^p.$$

For values of j that cannot be written in the form above we have $\mu_j = 0$. In particular for $j = 4^k$ we have $\mu_j = 1/2$.

The moments of μ provide the Taylor expansion of the Carathéodory function

$$F(z) = 1 + 2\sum_{j=1}^{\infty} \overline{\mu}_j z^j,$$

and from this it is not hard to compute the first few terms of the Taylor expansion of the Schur function $f(z)$ around $z = 0$. Indeed, from

$$F(z) = 1 + z^4 + \frac{z^{12}}{2} + z^{16} + \frac{z^{20}}{2} + \frac{z^{44}}{4} + \frac{z^{48}}{2} + \frac{z^{52}}{4} + \frac{z^{60}}{2} + z^{64} + \cdots$$

we get that $f(z)$ has the expansion

$$f(z) = \frac{z^3}{2} - \frac{z^7}{4} + \frac{3z^{11}}{8} + \frac{3z^{15}}{16} - \frac{z^{19}}{32} - \frac{5z^{23}}{64} - \frac{17z^{27}}{128} - \frac{29z^{31}}{256} + \cdots.$$

Only powers differing in multiples of 4 appear in the Taylor expansion of both functions, F and f. This follows from the fact that $d\mu(z) = d\nu(z^4)$ with ν given by the same infinite product (5.1) as μ but starting at $k = 0$, i.e. $d\nu(z) = (1 + (z + z^{-1})/2)\, d\mu(z)$. From this we find that $F(z) = G(z^4)$ and $f(z) = z^3 g(z^4)$ where G and g are the Carathéodory and Schur functions of ν respectively.

It is now possible, in principle, to compute as many Verblunsky parameters for the function g as one wishes; they are given by the continued fraction algorithm given at the end of section 5.2. The first few ones are given below, arranged for convenience in groups of eight. We list separately the first four parameters.

				1/2	−1/3	5/8	−1/13
1/14	−1/15	−1/4	−1/9	1/10	−1/11	21/32	−1/53
1/54	−1/55	−3/52	−1/49	1/50	−1/51	5/56	−1/61
1/62	−1/63	−1/20	−1/57	1/58	−1/59	−11/48	−1/37
1/38	−1/39	−1/12	−1/33	1/34	−1/35	1/8	−1/45
...							

It is clear that we get the Verblunsky parameters of f by introducing three zeros in between any two values above (a consequence of the argument z^4 in g above) and then shifting the resulting sequence by adding three extra zeros at the very beginning (a consequence of the factor z^3 in front of g above), yielding finally the following sequence of Verblunsky parameters for f, where each row contains eight coefficients starting with α_0 in the first row, α_8 in the second one, etc.

$$\begin{array}{cccrcccr}
0 & 0 & 0 & 1/2 & 0 & 0 & 0 & -1/3 \\
0 & 0 & 0 & 5/8 & 0 & 0 & 0 & -1/13 \\
0 & 0 & 0 & 1/14 & 0 & 0 & 0 & -1/15 \\
0 & 0 & 0 & -1/4 & 0 & 0 & 0 & -1/9 \\
0 & 0 & 0 & 1/10 & 0 & 0 & 0 & -1/11 \\
0 & 0 & 0 & 21/32 & 0 & 0 & 0 & -1/53 \\
0 & 0 & 0 & 1/54 & 0 & 0 & 0 & -1/55 \\
0 & 0 & 0 & -3/52 & 0 & 0 & 0 & -1/49 \\
0 & 0 & 0 & 1/50 & 0 & 0 & 0 & -1/51 \\
0 & 0 & 0 & 5/56 & 0 & 0 & 0 & -1/61 \\
\ldots
\end{array}$$

The non-zero Verblunsky parameters of f are given by

$$\xi_m \equiv \alpha_{4m-1}, \qquad m = 1, 2, 3, \ldots$$

where the sequence $\{\xi_m\}$ will be determined below. In fact it will be enough to determine the subsequence $\{\xi_{4+8n}\}$ since all the other values of ξ_m can be given by simple expressions in terms of these.

The expression for these $\xi_{4+8n} \equiv \alpha_{15+32n}$ is given by

$$\alpha_{15+32n} = -1/A_{n+1}, \qquad n = 0, 1, 2, \ldots$$

for a sequence of integer values $\{A_n\}$ to be described below. We will later give a different description of the complete sequence $\{\xi_m\}$ which makes clear what its limit points are and obviates the need to consider the subsequence $\{\xi_{4+8n}\}$.

We will first describe a procedure that allows us to generate the infinite sequence of integers $\{A_n\}_{n=1}^{\infty}$ of which the first ones are

$$13, 53, 61, 37, 45, 213, 221, 197, 205, 245, 253, 229, 237, 149, 157, 133, 141, \ldots$$

Once this sequence is accounted for, i.e. if the Verblunsky parameters of the form α_{15+32n} are known, we will see that all the remaining ones are determined by simple explicit formulas in terms of $\{A_n\}$. For this reason we will refer to the sequence $\{A_n\}$ to be constructed in the next section as the backbone of the sequence $\{\alpha_n\}$ we are interested in.

As we noted at the end of introduction, F. Riesz included the factor corresponding to $k = 0$ in the infinite product (5.1). That is, the measure considered by Riesz is the measure ν giving our measure μ (starting the infinite product with $k = 1$) by replacing z by z^4, and the corresponding Schur function is g. Therefore, the Verblunsky parameters $\{\alpha_n^R\}$ that F. Riesz would have are those obtained deleting in the sequence $\{\alpha_n\}$ the groups of three consecutive zeros, so $\alpha_n^R = \alpha_{4n+3} = \xi_{n+1}$, and all of them should be computed from $\alpha_{8n-5}^R = -1/A_n$.

The main difference between including the factor $k = 0$ in (5.1) or leaving it out is the inclusion of many zeros in the list of Verblunsky parameters in the second case, which is the one we choose. This makes for a much sparser CMV matrix which is easier to analyze than it would be in the original case of F. Riesz. On the other hand his choice is better for computational purposes when, of necessity, one has to deal with truncated matrices. In Riesz's case there is more information packed in the same size finite matrix. This point is exploited in some of the graphs displayed at the end of the paper.

5.6 Building the backbone

Consider the sets v_j, $j \geq 0$, defined as the ordered set of integers of the form

$$((-2)^j - 1)/3 + k2^{j+1}$$

where k runs over the integers. As an illustration we give a few elements of the sets $v_0, v_1, v_2, v_3, \ldots, v_{10}$, namely

$$\begin{aligned}
v_0 &= \ldots, -6, -4, -2, 0, 2, 4, 6, \ldots; \\
v_1 &= \ldots, -13, -9, -5, -1, 3, 7, 11, 15, \ldots; \\
v_2 &= \ldots, -23, -15, -7, 1, 9, 17, 25, \ldots; \\
v_3 &= \ldots, -35, -19, -3, 13, 29, 45, \ldots; \\
v_4 &= \ldots, -91, -59, -27, 5, 37, 69, 101, \ldots; \\
v_5 &= \ldots, -267, -203, -139, -75, -11, 53, 117, 181, \ldots; \\
v_6 &= \ldots, -491, -363, -235, -107, 21, 149, 277, 405, \ldots; \\
v_7 &= \ldots, -1067, -811, -555, -299, -43, 213, 469, 725, \ldots; \\
v_8 &= \ldots, -1963, -1451, -939, -427, 85, 597, 1109, \ldots; \\
v_9 &= \ldots, -2219, -1195, -171, 853, 1877, 2901, \ldots; \\
v_{10} &= \ldots, -5803, -3755, -1707, 341, 2389, 4437, \ldots.
\end{aligned}$$

These sets v_j , $j \geq 0$, are disjoint and their union gives all integers. A simple argument to prove this was kindly supplied by B. Poonen.

We observe that d_j, defined as the first positive element in the infinite set v_j (corresponding either to the choice $k = 0$ or $k = 1$) is given as follows: if $j = 0$ then $d_0 = 2$ otherwise, for $n \geq 1$ we have

$$d_j = ((-2)^j - 1)/3 + (1 - (-1)^j)2^j.$$

Define now, for $n \geq 4$,

$$c_n = 8 + ((-2)^{n-4} - 1)2^5/3$$

so that the values of $c_4, c_5, c_6, c_7, \ldots$ are given by

$$8, -24, 40, -88, 168, -344, 680, -1368, 2728, \ldots$$

a sequence whose first differences are given by

$$(-2)^{n+1}, \quad n = 4, 5, 6, \ldots.$$

For each pair j, n, $j \geq 0$, $n \geq 0$, define $w_{j,n}$ as the number of positive elements in the sequence $\{v_j\}$ that are not larger than n. Notice that $\sum_{j=0}^{\infty} w_{j,n} = n$ and that $w_{j,0}$ is zero for all j.

To be very explicit, we have for instance $w_{0,40} = 20$, $w_{1,40} = 10$, $w_{2,40} = 5$, $w_{3,40} = 2$, $w_{4,40} = 2$, $w_{5,40} = 0$, $w_{6,40} = 1$, and all values of $w_{j,40}$ after these ones vanish.

It is possible to give an expression for $w_{j,n}$ in terms of the sequence $\{d_j\}$ defined above. In fact one has

$$w_{j,n} = \left\lfloor \frac{n + 2^{j+1} - d_j}{2^{j+1}} \right\rfloor$$

where we use the notation $\lfloor x \rfloor$ to indicate the integer part of the quantity x.

We are finally ready to put all the pieces together and define, for $n \geq 0$,

$$u_n = \sum_{j=0}^{\infty} w_{j,n} c_{4+j}.$$

Notice that by definition this is a finite sum since, for a given n the expression $w_{j,n}$ vanishes if j is large enough.

The reader will have no difficulty verifying that we get

$$A_i = 13 + u_{i-1}, \qquad i \geq 1.$$

Recall that we have, for $n \geq 0$,

$$\alpha_{15+32n} = -\frac{1}{A_{n+1}}$$

and, as mentioned earlier, we will see how all other values of α_j can be determined from these ones.

5.7 Building the sequence from its backbone

We have observed already that the first non-zero Verblunsky parameters of our measure are given by

$$\alpha_3 = 1/2, \quad \alpha_7 = -1/3, \quad \alpha_{11} = 5/8, \quad \alpha_{15} = -1/13, \ \ldots .$$

We will see now that, for $i \geq 15$ and using the sequence $\{A_i\}$ built above, we have a way of computing the non-zero values of α_j. Start by observing that, after $\alpha_{15} = -1/A_1$ we get for values of j between $j = 16$ and $j = 47$ the following non-zero Verblunsky parameters:

$$\alpha_{19} = \frac{1}{1 + A_1}, \qquad \alpha_{23} = -\frac{1}{2 + A_1},$$

followed by

$$\alpha_{31} = -\frac{1}{A_1 - 4}, \qquad \alpha_{35} = \frac{1}{A_1 - 3}, \qquad \alpha_{39} = -\frac{1}{A_1 - 2},$$

and finally

$$\alpha_{47} = -\frac{1}{A_2}.$$

The reader will have noticed that we did not give a prescription for α_{27} or for α_{43}. This is done now:

$$\alpha_{27} = -\frac{3}{A_1 - 1}, \qquad \alpha_{43} = \frac{A_2 - A_1 + 2}{A_2 + A_1 - 2}.$$

We have seen that the non-zero values of α_j for j in between $j = 15$ and $j = 47$ are all obtained from the values of A_1 and A_2. We claim that exactly the same recipe apply for values of j in the range from $16(2p-1) - 1$ and $16(2p+1) - 1$, with $p \geq 1$, namely we set

$$\alpha_{16(2p-1)-1} = -\frac{1}{A_p}, \qquad \alpha_{16(2p+1)-1} = -\frac{1}{A_{p+1}}$$

and fill in the **SEVEN** non-zero values of α_j in between these two by using expressions that are extensions of the ones above, namely,

$$\alpha_{16(2p-1)+3} = \frac{1}{1+A_p}, \qquad \alpha_{16(2p-1)+7} = -\frac{1}{2+A_p},$$
$$\alpha_{16(2p-1)+15} = -\frac{1}{A_p-4}, \qquad \alpha_{16(2p-1)+19} = \frac{1}{A_p-3},$$
$$\alpha_{16(2p-1)+23} = -\frac{1}{A_p-2}, \qquad \alpha_{16(2p-1)+31} = -\frac{1}{A_{p+1}},$$

and, just as above, the missing Verblunsky parameters are given by

$$\alpha_{16(2p-1)+11} = -\frac{3}{A_p-1}, \qquad \alpha_{16(2p-1)+27} = \frac{A_{p+1}-A_p+2}{A_{p+1}+A_p-2}.$$

5.8 A different expression for the Verblunsky parameters

The construction above gives as many non-zero Verblunsky parameters for our measure as one wants, starting with $\alpha_3, \alpha_7, \alpha_{11}, \alpha_{15}, \ldots$ for which we get the values $1/2, 1/3, 5/8, -1/13, \ldots$.

In this section we give an explicit formula for these Verblunsky parameters in terms of a sequence of constants $\{K_i\}$, $i = 0, 1, 2, \ldots$, which are closely related to the sequence $\{A_i\}$ introduced above. One of the advantages of this new expression is that the set of limit values of the sequence $\{\alpha_j\}$ becomes obvious and is given by the union of three infinite sets, namely

$$-\frac{2}{K_i}, \qquad i = 1, 2, 3, \ldots,$$
$$\frac{4}{K_i+3}, \qquad i = 0, 1, 2, 3, \ldots,$$
$$-\frac{2}{K_i+6}, \qquad i = 0, 1, 2, 3, \ldots.$$

where the constants K_i are given by

$$K_0 = 3,$$
$$K_{2i-1} = 3A_i, \qquad K_{2i} = 3(A_i - 4), \qquad i = 1, 2, 3, \ldots.$$

One needs to add the limit points of the three infinite sets given above to get all limit points of the sequence $\{\alpha_j\}$.

We are now ready to give the alternative expressions for the non-zero Verblunsky parameters alluded to above, i.e. ξ_n so that

$$\xi_1 = 1/2, \quad \xi_2 = -1/3, \quad \xi_3 = 5/8, \ \dots,$$

and in general $\alpha_{4m-1} = \xi_m$. For this purpose we consider a disjoint union of the set of all non-negative integers into sets B_n, where the index n runs over the set $1, 2, 4, 5, 6, 8, \dots$, i.e. all positive $n \neq 3 \pmod 4$.

For each such n, define B_n as the set of integers of the form

$$\frac{1}{3} + 4^p \frac{3n-1}{3}, \qquad p = 0, 1, 2, 3, \dots.$$

Once again, a simple proof of these properties of the sets B_n was supplied by B. Poonen.

The sets B_n break naturally into three classes, with $n \equiv 0 \pmod 4$, $n \equiv 1 \pmod 4$ and $n \equiv 2 \pmod 4$. We claim that

$$\xi_{\frac{1}{3}+4^p\frac{3n-1}{3}} = -\frac{1}{K_s}\left(2 + \frac{1}{4^p}\right), \qquad n = 4s, \qquad s = 1, 2, 3, \dots,$$

$$\xi_{\frac{1}{3}+4^p\frac{3n-1}{3}} = \frac{1}{K_s + 3}\left(4 - \frac{1}{4^p}\right), \qquad n = 4s+1, \qquad s = 0, 1, 2, \dots,$$

$$\xi_{\frac{1}{3}+4^p\frac{3n-1}{3}} = -\frac{1}{K_s + 6}\left(2 + \frac{1}{4^p}\right), \qquad n = 4s+2, \qquad s = 0, 1, 2, \dots.$$

From the expressions above it follows that we have identified the limit values of the sequence $\{\alpha_{4m-1}\}$. The largest one is $2/3$ and the lowest one $-2/9$.

5.9 Some properties of the Riesz quantum walk

Once we get our hands on the Verblunsky parameters corresponding to the Riesz measure we construct the corresponding CMV matrix and we can compute different quantities pertaining to the associated quantum walk. In the rest of the paper we choose to illustrate some of these results with a few plots.

Figures 5.1 and 5.2 display the Verblunsky parameters themselves, key ingredients in the one-step transition amplitudes of the Riesz quantum walk. They show an apparent chaotic behaviour which is actually driven by the rules previously described which generate the full sequence of Verblunsky parameters. This is in great contrast to the translation invariant case of a constant coin.

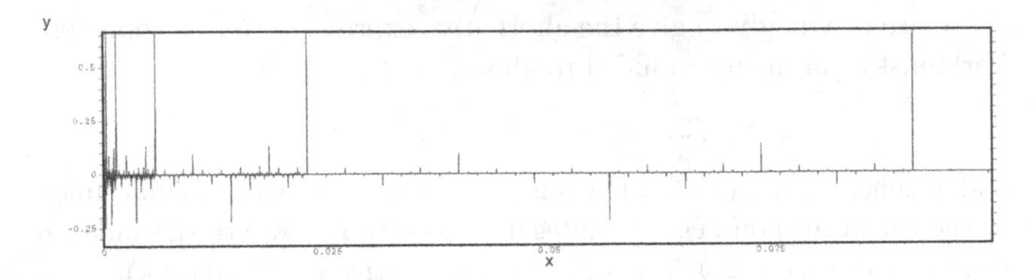

Figure 5.1 Riesz's measure: The first 30000 non-zero Verblunsky parameters.

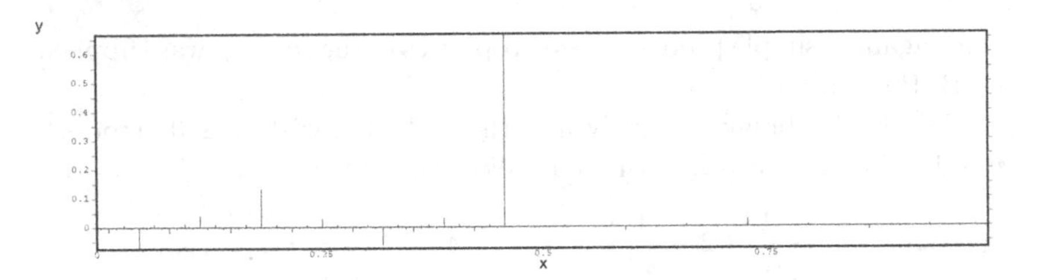

Figure 5.2 Riesz's measure: The non-zero Verblunsky parameters for indices between 30000 and 60000.

Figures 5.3 and 5.4 display the Taylor coefficients of the Schur function for Riesz measure and the Hadamard quantum walk with constant coin

$$C = \frac{1}{\sqrt{2}} \begin{pmatrix} 1 & 1 \\ 1 & -1 \end{pmatrix}$$

on the non-negative integers. These coefficients have an important probabilistic meaning discussed in great detail in the forthcoming paper [9]: the n-th Taylor coefficient is the first time return amplitude in n steps to the state with spin up at site 0.

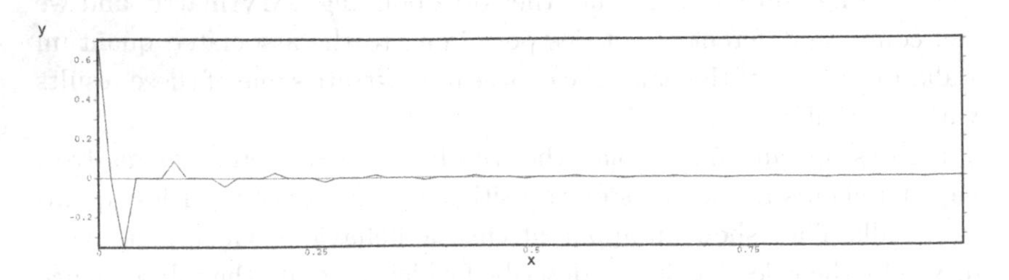

Figure 5.3 Hadmard's walk: The first 70 non-zero Taylor coefficients of the Schur function.

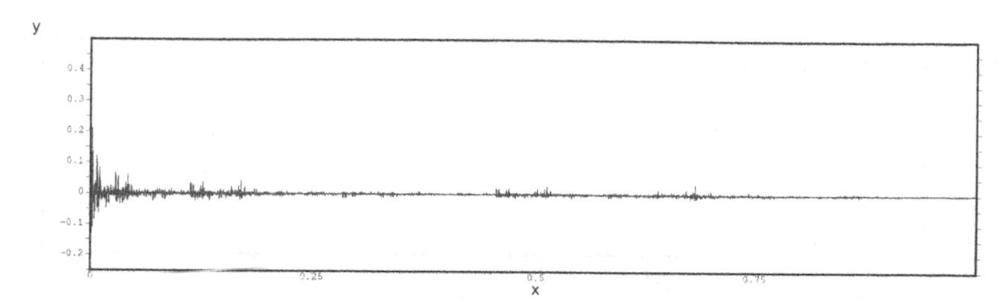

Figure 5.4 Riesz's measure: The first 7000 non-zero Taylor coefficients of the Schur function.

The first time return amplitudes for the Riesz walk seem to fluctuate in an apparent random way around a mean value which must decrease strongly enough to ensure that the sequence is square-summable, as Figure 5.5 makes evident. This is because the sum of the first return probabilities is the total return probability, which cannot be greater than one. Equivalently, any Schur function is Lebesgue integrable on the unit circle with norm bounded by one, and its norm is the sum of the squared moduli of the Taylor coefficients.

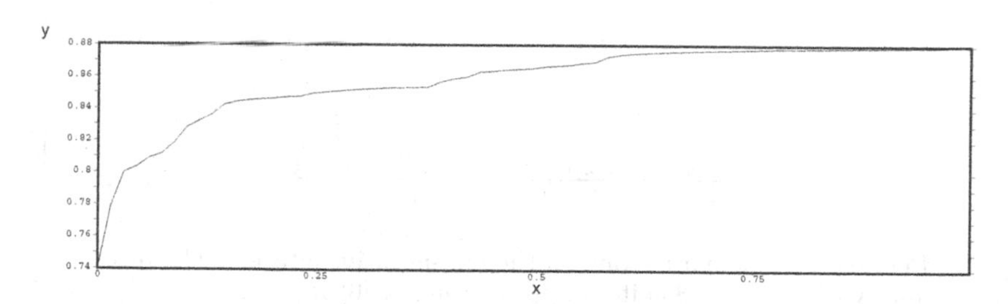

Figure 5.5 Riesz's measure: The cumulative sums of the squares of the first 7000 non-zero Taylor coefficients of the Schur function in steps of 100.

In the Hadamard case the behaviour of the first time return amplitudes is much more regular and the convergence to the total return probability, depicted in Figure 5.6, holds with a much higher speed. We should remark that the plot for the Hadamard walk picks up only the first 70 non-null coefficients, while the Riesz picture represents the first non-null 7000 coefficients. Hence, the differences between these two examples are not only in the more regular pattern that the Hadamard return probabilities exhibit, but also in the much higher tail for the Riesz return probabilities.

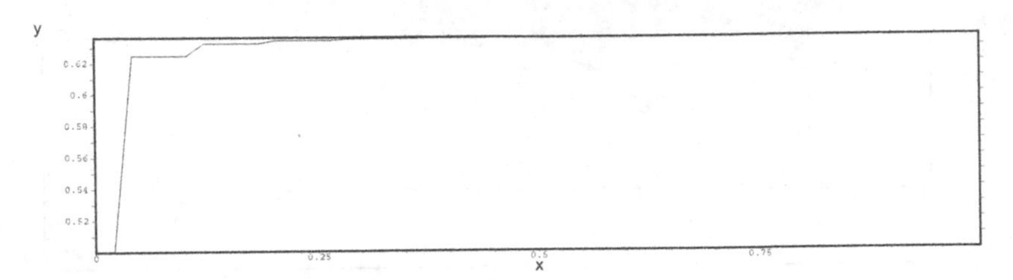

Figure 5.6 Hadamard's walk: The cumulative sums of the squares of the first 70 non-zero Taylor coefficients of the Schur function.

Figures 5.7 and 5.8 give the probability distribution of the random variable X_n/n, where X_n stands for the position (regardless of spin orientation) after n steps of the quantum walk started at position 0 with a spin pointing up. The plots given here correspond to the value $n = 800$ for both, the Riesz quantum walk and the Hadamard constant coin on the non-negative integers.

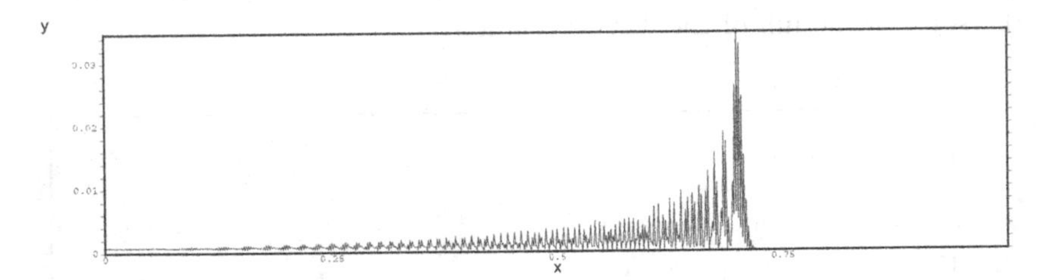

Figure 5.7 Hadamard's walk on the non-negative integers: The probability of X_n/n for 800 iterations starting at $|0\rangle \otimes |\uparrow\rangle$.

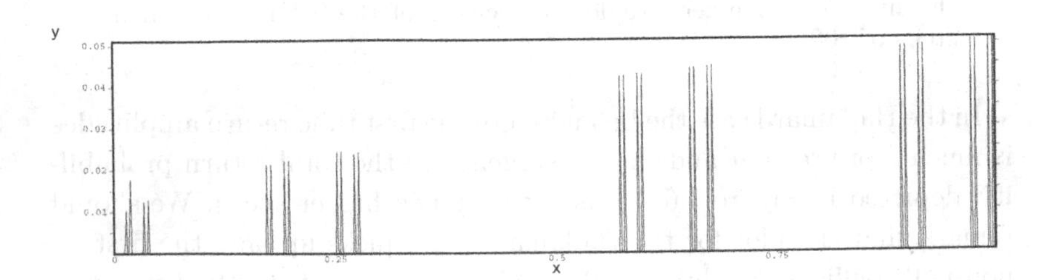

Figure 5.8 Riesz's walk on the non-negative integers: The probability of X_n/n for 800 iterations starting at $|0\rangle \otimes |\uparrow\rangle$.

The figures show that, in contrast to classical random walks for which

X_n behaves typically as $\sqrt{n}$, the position in a quantum walk can grow linearly with n. Nevertheless, Figure 5.8 shows a striking behaviour of the Riesz walk compared to the more regular asymptotics of the Hadamard walk reflected in Figure 5.7. This should be viewed as a clear indication of the anomalous behaviour that can appear under the presence of a singular continuous spectrum. In particular, these results make evident that quantum walks with a singular continuous measure can not exhibit nice limit laws as other toy models do. For the case of translation invariant ones it is known that obey inverted bell asymptotic distributions (see for instance [13]).

These results should motivate a more detailed analysis of quantum walks associated with singular continuous measures. This could lead to the discovery of new interesting quantum phenomena.

References

[1] A. Ambainis, Quantum walks and their algorithmic applications. *International Journal of Quantum Information*, **1**, (2003) 507–518.

[2] A. Ambainis, E. Bach, A. Nayak, A. Vishwanath and J. Watrous, One dimensional quantum walks. in *Proc. of the ACM Symposium on Theory and Computation (STOC'01)*, July 2001, ACM, NY, 2001, 37–49.

[3] O. Bourget, J. S. Howland, and A. Joye, Spectral analysis of unitary band matrices. *Commun. Math. Phys.*, **234**, (2003) 191–227.

[4] M. J. Cantero, L. Moral and L. Velázquez, Five-diagonal matrices and zeros of orthogonal polynomials on the unit circle. *Linear Algebra Appl.*, **362**, (2003) 29–56.

[5] M. J. Cantero, L. Moral and L. Velázquez, Minimal representations of unitary operators and orthogonal polynomials on the unit circle. *Linear Algebra Appl.*, **405**, (2005) 40–65.

[6] M. J. Cantero, F. A. Grünbaum, L. Moral and L. Velázquez, Matrix valued Szegő polynomials and quantum random walks. *Commun. Pure Applied Math.*, **58**, (2010) 464–507.

[7] Ya. L. Geronimus, On polynomials orthogonal on the circle, on trigonometric moment problem, and on allied Carathéodory and Schur functions. *Mat. Sb.*, **15**, (1944) 99–130. [Russian]

[8] C. G. Graham and O. C. McGehee, *Essays in Commutative Harmonic Analysis.* Springer-Verlag, 1979, Chapter 7.

[9] F. A. Grünbaum, L. Velazquez, A. Werner and R. Werner, Recurrence for discrete time unitary evolutions. In preparation.

[10] F. A. Grünbaum and L. Velazquez, The Riesz quantum walk on the integers. In preparation.

[11] Y. Katznelson, *An Introduction to Harmonic Analysis.* John Wiley & Sons, 1968.

[12] J. Kempe, Quantum random walks-an introductory overview. *Contemporary Physics*, **44**, (2003) 307–327.

[13] N. Konno, Quantum walks. in *Quantum Potential Theory*, U. Franz, M. Schürmann, editors, Lecture Notes in Mathematics 1954, Springer-Verlag, Berlin, Heidelberg, 2008.

[14] A. Magnus, Freund equation for Legendre polynomials on a circular arc and solution to the Grünbaum-Delsarte-Janssen-deVries problem. *J. Approx. Theory*, **139**, (2006) 75–90.

[15] D. Meyer, From quantum cellular automata to quantum lattice gases. *J. Stat. Physics*, **85**, (1996) 551–574, quant-ph/9604003.

[16] A. Nayak and A. Vishwanath, *Quantum walk on the line.* Center for Discrete Mathematics & Theoretical Computer Science, 2000, quant-ph/0010117.

[17] G. Pólya, Über eine Aufgabe der Wahrscheinlichkeitsrechnung betreffend die Irrfahrt im Strassennetz. *Math. Annalen*, **84**, (1921) 149–160.

[18] F. Riesz, Über die Fourierkoeffizienten einer stetigen Funktion von beschränkter Schwankung. *Math. Z.*, **18**, (1918) 312–315.

[19] F. Riesz and B. Sz-Nagy, *Functional Analysis.* F. Ungar Publishing, New York, 1955.

[20] W. Rudin, *Real and Complex Analysis*, second edition. McGraw-Hill Book Co., New York-Düsseldorf-Johannesburg, 1974.

[21] I. Schur, Über Potenzreihen die im Innern des Einheitskreises beschränkt sind. *J. Reine Angew. Math.*, **147**, (1916) 205–232 and **148**, (1917) 122–145.

[22] B. Simon, *Orthogonal Polynomials on the Unit Circle, Part 1: Classical Theory.* AMS Colloq. Publ., vol. 54.1, AMS, Providence, RI, 2005.

[23] B. Simon, CMV matrices: Five years after. *J. Comput. Appl. Math.*, **208**, (2007) 120–154.

[24] M. Stefanak, T. Kiss and I. Jex, Recurrence properties of unbiased coined quantum walks on infinite d dimensional lattices. arXiv: 0805.1322v2 [quant-ph] 4 Sep 2008.

[25] G. Szegő, *Orthogonal Polynomials*, 4th ed. AMS Colloq. Publ., vol. 23, AMS, Providence, RI, 1975.

[26] H. Wall, Continued fractions and bounded analytic functions. *Bull. Amer. Math. Soc.*, **50**, (1944) 110-119.

[27] D. S. Watkins, Some perspectives on the eigenvalue problem. *SIAM Rev.*, **35**, (1993) 430–471.

[28] A. Zygmund, *Trigonometric Series*, 2nd ed. Cambridge University Press, 1959.

6

Modulated Fourier Expansions for Continuous and Discrete Oscillatory Systems

Ernst Hairer
Section de Mathématiques
Université de Genève

Christian Lubich
Mathematisches Institut
Universität Tübingen

Abstract

This article reviews some of the phenomena and theoretical results on the long-time energy behaviour of continuous and discretized oscillatory systems that can be explained by modulated Fourier expansions: long-time preservation of total and oscillatory energies in oscillatory Hamiltonian systems and their numerical discretizations, near-conservation of energy and angular momentum of symmetric multistep methods for celestial mechanics, metastable energy strata in nonlinear wave equations. We describe what modulated Fourier expansions are and what they are good for.

6.1 Introduction

As a new analytical tool developed in the past decade, modulated Fourier expansions have been found useful to explain various long-time phenomena in both continuous and discretized oscillatory Hamiltonian systems, ordinary differential equations as well as partial differential equations. In addition, modulated Fourier expansions have turned out useful as a numerical approximation method in oscillatory systems.

In this review paper we first show some long-time phenomena in oscillatory systems, then give theoretical results that explain these phenomena, and finally outline the basics of modulated Fourier expansions with which these results are proved.

6.2 Some phenomena

6.2.1 Time scales in a nonlinear oscillator chain

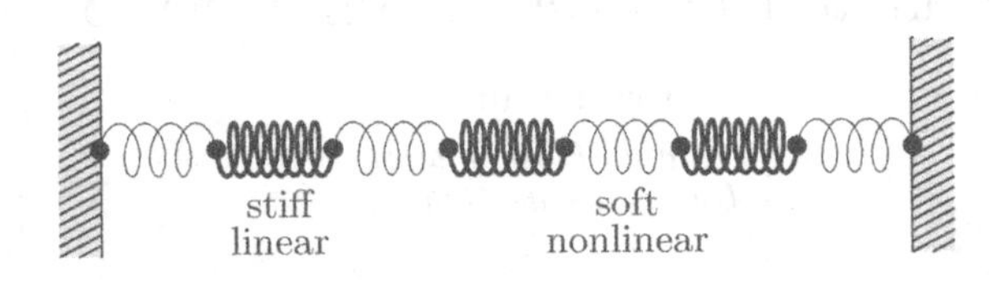

Figure 6.1 Particle chain with alternating soft nonlinear and stiff linear springs.

Following Galgani, Giorgilli, Martinoli & Vanzini [13], we consider a chain of particles interconnected alternately by stiff linear springs and soft nonlinear springs, as shown in Figure 6.1. We assume that the particles are of unit mass and that the spring constant of the stiff linear springs is ε^{-2} for a small parameter ε. This example was chosen as a model problem for nonlinear oscillatory Hamiltonian problems with a single constant high frequency $\omega = 1/\varepsilon$ in [25, Chap. XIII].

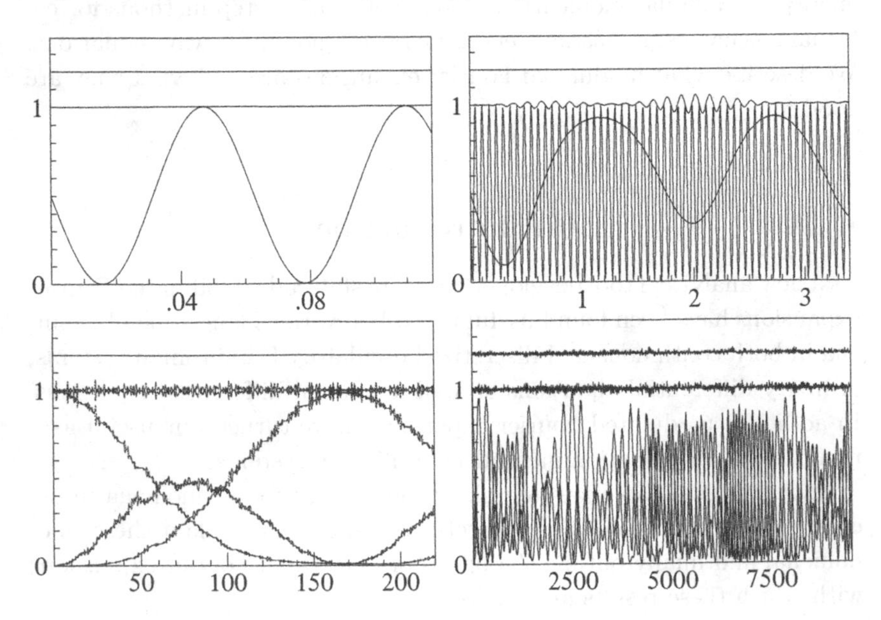

Figure 6.2 Different time scales in the oscillator chain ($\varepsilon = 1/50$).

The system shows different behaviour on several time scales. On the fast time scale ε there is the almost-sinusoidal vibration of the stiff springs. The time scale ε^0 is that of the motion of the nonlinear springs. On the slow time scale ε^{-1}, there is an energy transfer between the stiff springs. Over very long times ε^{-N} with $N > 1$ and even over exponentially long times in $1/\varepsilon$ ($t \le e^{c/\varepsilon}$), there is almost-conservation of the sum of the harmonic energies of the stiff springs. The total energy is conserved for all times.

This behaviour is illustrated in Figure 6.2, where various energies in the system are plotted as functions of time. In all four pictures, the upper constant line is the total energy. In the final, long-time picture the total energy does not appear as constant, because the upper line represents the total energy along a numerical solution that was computed with large step size $h > \varepsilon$. It is remarkable that the numerical method, which here is a trigonometric integrator, shows no drift in the total energy over such long times, just small oscillations about the correct value. The lower almost-constant line in the pictures represents the oscillatory energy, that is, the sum of the harmonic energies of the stiff springs. It is not exactly conserved, but stays close to its initial value over extremely long times, both in the exact and the numerical solution. This calls for an explanation.

Let us briefly describe the further curves in Figure 6.2: the sinusoidal curve in the first picture represents the kinetic energy of the first stiff spring, the smooth curve in the second picture represents the kinetic energy in the first nonlinear spring. The third and fourth picture show the energy transfer between the stiff springs: the curves show the harmonic energies of the three stiff springs, starting from an initial configuration where only the first spring is excited, while the second and third stiff springs are at rest.

6.2.2 Symmetric multistep methods over long times

We consider the numerical solution of second-order differential equations

$$\ddot{x} = -\nabla U(x)$$

by linear multistep methods. It was reported in the astrophysical literature [26] that some *symmetric* multistep methods exhibit excellent long-time behaviour in the computation of planetary orbits, similar to that known for symplectic one-step methods. Since multistep methods cannot be symplectic, as was shown by Tang [28], such behaviour comes unexpected.

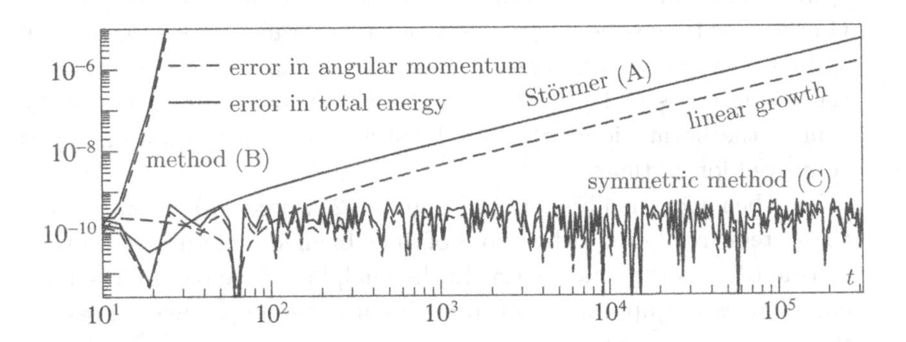

Figure 6.3 Error in energy and angular momentum of three multistep methods applied to the Kepler problem.

Figure 6.3 shows the error in the total energy and the angular momentum of three multistep methods applied to the Kepler problem, all three of the same order 8. The linear error growth in energy and angular momentum corresponds to a non-symmetric method, the eighth-order Störmer method (A). One symmetric method exhibits exponential error growth (B), but another symmetric method shows no drift in energy and angular momentum (C). Such behaviour needs to be explained.

6.2.3 Metastable energy strata in nonlinear wave equations

We consider the nonlinear wave equation $u_{tt} - u_{xx} + \frac{1}{2}u = u^2$ with periodic boundary conditions on a space interval of length 2π and report a numerical experiment from [17]. The initial data are chosen such that only the first Fourier mode is excited initially, with harmonic energy $E_1(0) = \varepsilon = 10^{-4}$. All higher-mode harmonic energies are initially zero. As is shown in Figure 6.4, due to the presence of the nonlinearity, they become non-zero immediately. Surprisingly, however, the jth mode energy settles quickly at level ε^j (the zero-mode energy at ε^2) and stays there for extremely long times. There is no perceptible energy transfer among the modes.

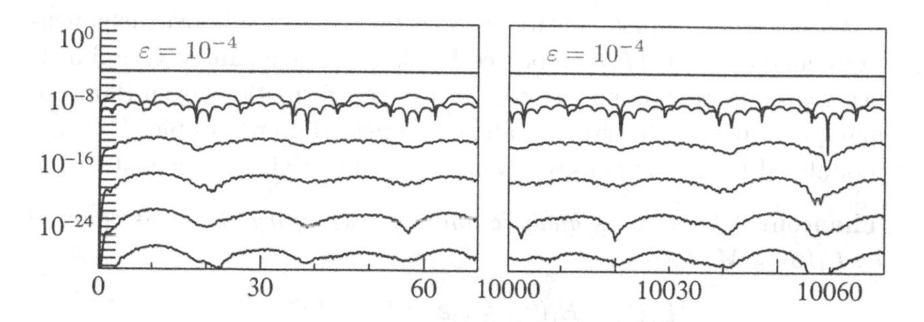

Figure 6.4 Mode energies versus time on different time windows.

6.3 Some theorems

We give theoretical results that explain the long-time phenomena encountered in Section 6.2.

6.3.1 Oscillatory ordinary differential equations

We consider a system of second-order differential equations for $x_0 \in \mathbb{R}^{d_0}$, $x_1 \in \mathbb{R}^{d_1}$,

$$\begin{aligned} \ddot{x}_0 &= -\nabla_{x_0} U(x_0, x_1) \\ \ddot{x}_1 + \frac{1}{\varepsilon^2} x_1 &= -\nabla_{x_1} U(x_0, x_1), \qquad 0 < \varepsilon \ll 1. \end{aligned} \tag{6.1}$$

With the momenta $y_0 = \dot{x}_0$, $y_1 = \dot{x}_1$, this is a Hamiltonian system with the Hamilton function

$$H(x_0, x_1, y_0, y_1) = \frac{1}{2}\,\|y_0\|^2 + \frac{1}{2}\,\|y_1\|^2 + \frac{1}{2}\,\varepsilon^{-2}\|x_1\|^2 + U(x_0, x_1).$$

Example. If we describe the position of the $2m$ particles in the nonlinear oscillator chain of Figure 6.1 (where $m = 3$) by the coordinates of the centres of the stiff linear springs, $x_0 = (x_{0,1}, \dots, x_{0,m})$, and by their elongations from the rest length, $x_1 = (x_{1,1}, \dots, x_{1,m})$, then the equations of motion take the above form. The potential U is given as $U(x_0, x_1) = V(s_0) + \dots + V(s_m)$, where s_j denotes the elongation of the jth soft nonlinear spring with potential V.

The oscillatory energy

$$E_1 = \frac{1}{2}\,\|\dot{x}_1\|^2 + \frac{1}{2}\,\varepsilon^{-2}\|x_1\|^2 = \sum_{j=1}^{d_1} \left(\frac{1}{2}\,\dot{x}_{1,j}^2 + \frac{1}{2}\,\varepsilon^{-2} x_{1,j}^2 \right)$$

turns out to be almost invariant. The following result over exponentially long times in $1/\varepsilon$ was proved by Benettin, Galgani & Giorgilli [1] using a sequence of nonlinear coordinate transformations from Hamiltonian perturbation theory, and later by Cohen, Hairer & Lubich [4] using modulated Fourier expansions, working in the original coordinates.

Theorem 6.1 *If U is analytic and the oscillatory energy is bounded by $E_1(0) \le M$, then*

$$|E_1(t) - E_1(0)| \le C\varepsilon \quad \textit{for} \quad t \le e^{c/\varepsilon},$$

provided that $(x_0(t), 0)$ stays in a compact subset of the domain of analyticity of U. The constants C and c are independent of ε, but depend on M, $\|\dot{x}_0(0)\|$ and bounds of U.

The condition on $x_0(t)$ is satisfied, for example, if $U(x_0, 0) \to +\infty$ as $\|x_0\| \to \infty$, since the total energy H is constant along the solution.

Theorem 6.1 explains the almost constant line for E_1 in the lower right picture of Figure 6.2 — at least for the exact solution, whereas the picture was obtained with a numerical method with large step size $h > \varepsilon$, which nevertheless shows remarkably good long-time energy behaviour. The numerical method employed is a *trigonometric integrator*, a method that is exact for $\ddot{x}_1 + \varepsilon^{-2} x_1 = 0$ and reduces to the Störmer-Verlet method for $\ddot{x}_0 = f(x_0)$. The method is of the form

$$\begin{aligned}
x_0^{n+1} &= x_0^n + h\dot{x}_0^n + \tfrac{1}{2} h^2 g_0^n \\
\dot{x}_0^{n+1} &= \dot{x}_0^n + \tfrac{1}{2} h \big(g_0^n + g_0^{n+1}\big) \\[2ex]
x_1^{n+1} &= \cos\Big(\frac{h}{\varepsilon}\Big)\, x_1^n + \varepsilon \sin\Big(\frac{h}{\varepsilon}\Big)\, \dot{x}_1^n + \tfrac{1}{2} h^2 \psi\Big(\frac{h}{\varepsilon}\Big)\, g_1^n \\
\dot{x}_1^{n+1} &= -\frac{1}{\varepsilon} \sin\Big(\frac{h}{\varepsilon}\Big)\, x_1^n + \cos\Big(\frac{h}{\varepsilon}\Big)\, \dot{x}_1^n + \tfrac{h}{2} \Big(\psi_0\Big(\frac{h}{\varepsilon}\Big)\, g_1^n + \psi_1\Big(\frac{h}{\varepsilon}\Big)\, g_1^{n+1}\Big)
\end{aligned}$$

with $g_j^n = -\nabla_{x_j} U\big(x_0^n, \phi\big(\tfrac{h}{\varepsilon}\big) x_1^n\big)$ for $j = 0, 1$ and with filter functions ψ, ψ_0, ψ_1, ϕ that take the value 1 at 0. Exchanging $n \leftrightarrow n+1$ and $h \leftrightarrow -h$ in the method, it is seen that the method is symmetric for all values of h and ε if and only if $\psi(\xi) = \operatorname{sinc}(\xi)\, \psi_1(\xi)$ (where $\operatorname{sinc}(\xi) = \sin(\xi)/\xi$) and $\psi_0(\xi) = \cos(\xi)\, \psi_1(\xi)$, which we assume in the following. The method is symplectic if and only if $\psi(\xi) = \operatorname{sinc}(\xi)\, \phi(\xi)$. A popular choice, proposed by Garcia-Archilla, Sanz-Serna & Skeel [15], is $\phi(\xi) = \operatorname{sinc}(\xi)$, $\psi(\xi) = \operatorname{sinc}(\xi)^2$, but other choices are also favoured in the literature, see Hairer, Lubich & Wanner [25, Chap. XIII] and Grimm & Hochbruck [20].

We are interested in the long-time behaviour of the total energy

$$H^n = H(x_0^n, x_1^n, \dot{x}_0^n, \dot{x}_1^n)$$

and oscillatory energy

$$E_1^n = \frac{1}{2}\,\|\dot{x}_1^n\|^2 + \frac{1}{2}\,\varepsilon^{-2}\|x_1^n\|^2$$

along the numerical solution. For a numerical analogue of Theorem 6.1 we need the following conditions:

- the energy bound $E_1^0 \le M$
- a condition on the numerical solution: the values $(x_0^n, 0)$ stay in a compact subset of a domain on which the potential U is smooth;
- conditions on the filter functions: ψ and ϕ have no real zeros other than integral multiples of π; they satisfy

$$|\psi(h/\varepsilon)| \le C_1 \operatorname{sinc}^2(\tfrac{1}{2}\,h/\varepsilon)\,, \qquad |\phi(h/\varepsilon)| \le C_2\,|\operatorname{sinc}(\tfrac{1}{2}\,h/\varepsilon)|\,,$$
$$|\psi(h/\varepsilon)\phi(h/\varepsilon)| \le C_3\,|\operatorname{sinc}(h/\varepsilon)|\,;$$

- the condition $h/\varepsilon \ge c_0 > 0$;
- a numerical non-resonance condition: for some $N \ge 2$,

$$|\sin(\tfrac{1}{2}\,kh/\varepsilon)| \ge c\sqrt{h} \quad \text{for} \quad k = 1, \dots, N.$$

Theorem 6.2 *Under the above conditions, the total and oscillatory energies along the numerical solution satisfy*

$$\begin{aligned} H^n &= H^0 + \mathcal{O}(h) \\ E_1^n &= E_1^0 + \mathcal{O}(h) \end{aligned} \qquad \text{for} \quad 0 \le nh \le h^{-N+1}.$$

The constants symbolized by $\mathcal{O}$ are independent of n, h, ε satisfying the above conditions, but depend on N and the constants in the conditions.

This result from Hairer & Lubich [21], see also Hairer, Lubich & Wanner [25, Chap. XIII], was the first long-time result proved with modulated Fourier expansions. That technique could easily be transferred from the continuous to the discrete problem, which does not seem possible for the nonlinear coordinate transforms of Hamiltonian perturbation theory with which Theorem 6.1 was first proved.

The differential equation (6.1) has just a single high frequency $\omega = 1/\varepsilon$. Extensions of Theorem 6.1 to problems with several high frequencies are proved by Benettin, Galgani and Giorgilli [2] (via Hamiltonian perturbation theory) and Cohen, Hairer & Lubich [5] (via modulated Fourier expansions). The latter paper gives a multi-frequency version of Theorem 6.2 for the numerical solution. An additional difficulty in the multi-frequency case appears because of possible resonances among the frequencies, which need to be studied carefully. In the case of non-resonant frequencies $\omega_j = \lambda_j/\varepsilon$, each of the oscillatory energies $E_j = \frac{1}{2}\|\dot{x}_j\|^2 + \frac{1}{2}\omega_j^2\|x_j\|^2$ is nearly preserved over long times, but in resonant cases only certain linear combinations of the E_j, and in particular their sum, are nearly preserved.

6.3.2 Symmetric multistep methods over long times

We now consider a second-order system of ordinary differential equations with conservative forces (here with just one time scale)

$$\ddot{x} = f(x), \qquad f(x) = -\nabla U(x), \tag{6.2}$$

with a smooth potential U. The system is Hamiltonian with

$$H(x, \dot{x}) = \frac{1}{2}\,\|\dot{x}\|^2 + U(x).$$

For the numerical approximation $x^n \approx x(nh)$ we consider a linear multistep method with step size h,

$$\sum_{j=0}^{k} \alpha_j x^{n+j} = h^2 \sum_{j=0}^{k} \beta_j f(x^{n+j}), \tag{6.3}$$

with the following properties:

- the method is symmetric: $\alpha_j = \alpha_{k-j}$, $\beta_j = \beta_{k-j}$
- all zeros of $\sum \alpha_j \zeta^j$ lie on the unit circle and are simple, except a double root at 1
- the method is of order $p \geq 2$.

We recall that the order of the method is characterized by the relation

$$\frac{1}{h^2}\frac{\sum \alpha_j e^{jh}}{\sum \beta_j e^{jh}} = 1 + \mathcal{O}(h^p) \quad \text{for} \quad h \to 0,$$

which entails a double zero of $\sum \alpha_j \zeta^j$ at 1. The method is completed with a velocity approximation

$$\dot{x}^n \approx \frac{1}{h} \sum_{j=-\ell}^{\ell} \delta_j x^{n+j},$$

which we also assume to be of order p. The velocity approximations are computed *a posteriori* and are not propagated in the scheme. Moreover, we assume that the error of the starting approximations $x^0, \dots, x^{k-1}$ is $\mathcal{O}(h^{p+1})$.

The following result on the long-time near-conservation of the total energy $H^n = H(x^n, \dot{x}^n)$ is proved in [22].

Theorem 6.3 *Under the above assumptions,*

$$H^n = H^0 + \mathcal{O}(h^p) \quad \textit{for} \quad nh \le h^{-p-2}.$$

The constant symbolized by the $\mathcal{O}$-notation is independent of n and h with nh in the stated interval. If no root of $\sum \alpha_j \zeta^j$ other than 1 is a product of two other roots, then there is energy conservation up to $\mathcal{O}(h^p)$ even over times $nh \le h^{-2p-3}$.

Systems with a rotational symmetry preserve angular momentum. This comes about through an invariance

$$U(e^{\tau A}x) = U(x) \quad \text{for all } x \text{ and real } \tau$$

with a skew-symmetric matrix A via Noether's theorem: the system then has the quadratic first integral

$$L(x, \dot{x}) = \dot{x}^T A x.$$

The following result from [22] states the long-time near-conservation of angular momentum $L^n = L(x^n, \dot{x}^n)$ along the numerical solution.

Theorem 6.4 *Under the above assumptions,*

$$L^n = L^0 + \mathcal{O}(h^p) \quad \textit{for} \quad nh \le h^{-p-2}.$$

The constant symbolized by the $\mathcal{O}$-notation is independent of n and h with nh in the stated interval.

Theorems 6.3 and 6.4 explain the excellent long-time behaviour of the favourable of the three methods of Figure 6.3. The method with the linear error growth in energy and angular momentum is not symmetric, and the method with the exponential error growth is symmetric but its characteristic polynomial has further double roots apart from 1.

At first sight, the problem considered in this subsection bears little resemblance to that of the previous subsection, but in fact both can be viewed as perturbed linear problems. Here, it is the numerical scheme that introduces a fast time scale h on which the solutions of the linear recurrence relation

$$\sum_{j=0}^{k} \alpha_j x^{n+j} = 0$$

oscillate, with parasitic terms ζ_i^n, where the ζ_i are the zeros of the characteristic polynomial $\sum \alpha_j \zeta^j$ that are different from the principal root 1. The squared norms of parasitic solution components in the nonlinear problem play a similar role to the oscillatory energies E_i of the previous subsection and can be shown to remain almost constant along the numerical solution. The proof uses similar analytical techniques, that is, modulated Fourier expansions.

Bounding parasitic components is the main obstacle to obtaining similarly good long-time results for symmetric multistep methods applied to more general Hamiltonian systems than (6.2), see [25, Chap. XV] and Console & Hairer [10].

6.3.3 Metastable energy strata in nonlinear wave equations

We now turn to a surprising long-time result in weakly nonlinear partial differential equations. The *linear* Klein–Gordon equation

$$u_{tt} - \Delta u + \rho u = 0 \quad (x \in \mathbb{R}^d, t \in \mathbb{R}); \qquad \text{with} \quad \rho \geq 0$$

with initial data $a\, e^{ik\cdot x} + b\, e^{-ik\cdot x}$ for some (fixed) wave vector $k \in \mathbb{R}^d$ has a solution that is a linear combination of plane waves $e^{i(\pm k\cdot x \pm \omega t)}$ (with frequency $\omega = \sqrt{|k|^2 + \rho}$). We now consider the nonlinearly perturbed equation

$$u_{tt} - \Delta u + \rho u = g(u) \tag{6.4}$$

with the same initial data and ask the question: *Are plane waves stable under nonlinear perturbations of the equation?*

The solution has a Fourier series

$$u(x,t) = \sum_{j\in\mathbb{Z}} u_j(t)\, e^{ijk\cdot x},$$

where the coefficient functions satisfy the infinite system of second-order differential equations

$$\ddot{u}_j + \omega_j^2 u_j = -\frac{\partial U}{\partial u_{-j}}(\mathbf{u}), \qquad j \in \mathbb{Z},$$

with frequencies $\omega_j = \sqrt{j^2|k|^2 + \rho}$ and the potential

$$U(\mathbf{u}) = -\sum_m \frac{g^{(m-1)}(0)}{m!} \sum_{j_1+\cdots+j_m=0} u_{j_1}\cdots u_{j_m} \quad \text{for} \quad \mathbf{u} = (u_j)_{j\in\mathbb{Z}}.$$

We are interested in the size of the mode energies

$$E_j(t) = \frac{1}{2}|\dot{u}_j(t)|^2 + \frac{1}{2}|\omega_j u_j(t)|^2$$

for large t and in the energy transfer to higher modes.

We note that $E_j(t) = E_{-j}(t)$ for real solutions of (6.4). We assume the following:

- real initial data with $E_1(0) = \varepsilon$, $E_j(0) = 0$ for $|j| \neq 1$
- a real-analytic nonlinearity g with $g(0) = g'(0) = 0$.

The following long-term stability result is proved by Gauckler, Hairer, Lubich & Weiss [17] using modulated Fourier expansions in time.

Theorem 6.5 *Assume the above conditions and fix an integer $K > 1$. Then the following holds true: For almost all mass parameters $\rho > 0$ and wave vectors k, solutions to the nonlinear Klein–Gordon equation (6.4) satisfy, for sufficiently small ε and over long times*

$$t \leq c\varepsilon^{-K/4},$$

the bounds $|E_1(t) - E_1(0)| \leq C\varepsilon^2, \quad E_0(t) \leq C\varepsilon^2,$

$$E_j(t) \leq C\varepsilon^j, \quad 0 < j < K,$$

$$\sum_{j=K}^{\infty} \varepsilon^{-(j-K)/2} E_j(t) \leq C\varepsilon^K.$$

The constants C are independent of ε and t in the stated interval.

This result explains the stable energy strata observed in Figure 6.4. It holds for almost all (with respect to Lebesgue measure) $\rho > 0$ and k, for which it can be shown that the frequencies $\omega_j = \sqrt{j^2|k|^2 + \rho}$ satisfy a non-resonance condition. For $\rho = 0$ the frequencies are fully resonant, and there are no stable energy strata.

6.3.4 Further results

Modulated Fourier expansions have been used to prove a variety of further long-time results for oscillatory ordinary and partial differential equations and their numerical discretizations:

- energy distribution in Fermi–Pasta–Ulam chains of particles [24]
- long-time Sobolev regularity of nonlinear wave equations [6]
- Sobolev stability of plane wave solutions to nonlinear Schrödinger equations [11]
- long-time near-conservation of actions in Hamiltonian partial differential equations [6, 18, 16]
- ... and their numerical counterparts [7, 18, 19, 16, 23].

A common theme in all these works is the long-time behaviour of nonlinearly perturbed oscillatory systems and their numerical discretizations.

6.4 Modulated Fourier expansions

Modulated Fourier expansions in time have been developed as a technique for analysing weakly nonlinear oscillatory systems, both continuous and discrete, over long times. There are two ingredients:

- a solution approximation over short time (the modulated Fourier expansion properly speaking)
- almost-invariants of the modulation system.

Together they yield long-time results as illustrated in the previous section. The technique can be viewed as embedding the original system in a larger modulation system that turns out to have a Hamiltonian/Lagrangian structure with an invariance property from which the long-time behaviour can be inferred.

The technique was first developed for the long-time analysis of numerical integrators for highly oscillatory ordinary differential equations in [21] and was subsequently extended by the authors of the present review together with Cohen, Console, Gauckler and Weiss (see references) to treat analytical and numerical problems in Hamiltonian ODEs, PDEs, and lattice systems over long times. The approach was also taken up by Sanz-Serna [27] for analysing the heterogeneous multiscale method for oscillatory ODEs. In addition to the use of modulated Fourier expansions as an analytical technique, they have also been found useful as a numerical approximation method by Hairer, Lubich & Wanner [25, Chap. XIII], Cohen [3], and Condon, Deaño & Iserles [8, 9].

We illustrate the procedure on the model problem

$$\ddot{x}_j + \omega_j^2 x_j = \sum_{j_1+j_2=j \bmod N} x_{j_1} x_{j_2} \quad \text{for} \quad j = 1, \dots, N \tag{6.5}$$

for large frequencies $\omega_j = \lambda_j/\varepsilon$, with $\lambda_j \geq 1$. We assume that the oscillatory energies $E_j = \frac{1}{2}\dot{x}_j^2 + \frac{1}{2}\omega_j^2 x_j^2$ are initially bounded independently of ε.

We make the approximation ansatz

$$x_j(t) \approx \sum_{\mathbf{k}} z_j^{\mathbf{k}}(t)\, e^{i(\mathbf{k}\cdot\boldsymbol{\omega})t} \tag{6.6}$$

with slowly varying modulation functions $z_j^{\mathbf{k}}$, all derivatives of which should be bounded independently of ε. The sum is taken over a finite set of multi-indices $\mathbf{k} = (k_1, \ldots, k_N) \in \mathbb{Z}^N$, and $\mathbf{k}\cdot\boldsymbol{\omega} = \sum k_j\omega_j$. The slowly changing modulation functions appear multiplied with the highly oscillatory exponentials $e^{i(\mathbf{k}\cdot\boldsymbol{\omega})t} = \prod_{j=1}^N \big(e^{i\omega_j t}\big)^{k_j}$, which are products of solutions to the linear equations $\ddot{x}_j + \omega_j^2 x_j = 0$.

Modulation system and non-resonance condition. When we insert this ansatz into the differential equation (6.5) and collect the coefficients corresponding to the same exponential $e^{i(\mathbf{k}\cdot\boldsymbol{\omega})t}$, we obtain the infinite system of modulation equations

$$\big(\omega_j^2 - (\mathbf{k}\cdot\boldsymbol{\omega})^2\big)\, z_j^{\mathbf{k}} + 2i(\mathbf{k}\cdot\boldsymbol{\omega})\dot{z}_j^{\mathbf{k}} + \ddot{z}_j^{\mathbf{k}} = -\frac{\partial \mathcal{U}}{\partial z_{-j}^{-\mathbf{k}}}(\mathbf{z}). \tag{6.7}$$

The left-hand side results from the linear part in (6.5) and the right-hand side from the nonlinearity. It turns out to have a gradient structure with the modulation potential

$$\mathcal{U}(\mathbf{z}) = -\frac{1}{3} \sum_{j_1+j_2+j_3=0 \bmod N} \; \sum_{\mathbf{k}^1+\mathbf{k}^2+\mathbf{k}^3=0} z_{j_1}^{\mathbf{k}^1} z_{j_2}^{\mathbf{k}^2} z_{j_3}^{\mathbf{k}^3}.$$

The infinite system is truncated and can be solved approximately (up to a defect ε^K) for polynomial modulation functions $z_j^{\mathbf{k}}$ under a non-resonance condition: we require that small denominators $\omega_j^2 - (\mathbf{k}\cdot\boldsymbol{\omega})^2$ are not too small. For example, we might suppose, as in [5],

$$|\mathbf{k}\cdot\boldsymbol{\lambda}| \geq c\sqrt{\varepsilon} \quad \text{for} \quad \mathbf{k} \in \mathbb{Z}^N \quad \text{with} \quad 0 < |\mathbf{k}| \leq 2K,$$

where $|\mathbf{k}| = |k_1| + \cdots + |k_N|$. Under such a non-resonance condition one can construct and suitably bound the modulation functions $z_j^{\mathbf{k}}$, and the modulated Fourier expansion (6.6) is an $\mathcal{O}(\varepsilon^K)$ approximation to the solution over a short time interval $t = \mathcal{O}(1)$.

The case of resonant frequencies can also be treated by requiring a non-resonance condition outside a resonance module; cf. [2, 5]. The situation is at present less clear for almost-resonances.

Almost-invariants of the modulation system. With the functions $y_j^{\mathbf{k}}(t) = z_j^{\mathbf{k}}(t)e^{i(\mathbf{k}\cdot\boldsymbol{\omega})t}$, the modulation equations (6.7) take the Newtonian form

$$\ddot{y}_j^{\mathbf{k}} + \omega_j^2 y_j^{\mathbf{k}} = -\frac{\partial \mathcal{U}}{\partial y_{-j}^{-\mathbf{k}}}(\mathbf{y}).$$

The modulation potential has the obvious, but important invariance property

$$\mathcal{U}(S_\ell(\theta)\mathbf{z}) = \mathcal{U}(\mathbf{z}) \quad \text{for} \quad S_\ell(\theta)\mathbf{z} = (e^{ik_\ell\theta} z_j^{\mathbf{k}})_{j,\mathbf{k}}.$$

Formally applying Noether's theorem, this leads to formal invariants

$$\mathcal{E}_\ell(\mathbf{z}, \dot{\mathbf{z}}) = \frac{1}{2}\sum_j \sum_{\mathbf{k}} k_\ell \omega_\ell \Big((\mathbf{k}\cdot\boldsymbol{\omega})|z_j^{\mathbf{k}}|^2 - i z_{-j}^{-\mathbf{k}} \dot{z}_j^{\mathbf{k}} \Big),$$

which are almost-invariants of the truncated modulation system (6.7). They turn out to be close to the oscillatory energies E_ℓ. By patching together many short time intervals, the drift in the almost-invariants $\mathcal{E}_\ell$ can be controlled to remain small over long times, and in this way also the drift in the oscillatory energies E_ℓ is under control.

With these ingredients and many problem-specific technical details and estimates we obtain results on the long-time behaviour of the oscillatory energies E_ℓ.

6.5 Conclusion

We close this review with a citation from the very influential paper by Fermi, Pasta & Ulam [12] (see [14] for a reprint and review):

"This report is intended to be the first one in a series dealing with the behavior of certain nonlinear physical systems where the non-linearity is introduced as a perturbation to a primarily linear problem. The behavior of the systems is to be studied for times which are long compared to the characteristic periods of the corresponding linear problem."

This is just what modulated Fourier expansions are good for, both for continuous problems and their numerical discretizations.

References

[1] G. Benettin, L. Galgani and A. Giorgilli, Realization of holonomic constraints and freezing of high frequency degrees of freedom in the light of

classical perturbation theory. Part I. *Comm. Math. Phys.*, **113**, 87–103, 1987.

[2] G. Benettin, L. Galgani and A. Giorgilli Realization of holonomic constraints and freezing of high frequency degrees of freedom in the light of classical perturbation theory. Part II. *Comm. Math. Phys.*, **121**, 557–601, 1989.

[3] D. Cohen, *Analysis and numerical treatment of highly oscillatory differential equations.* Ph. D. Thesis, Univ. Genève, 2004.

[4] D. Cohen, E. Hairer and C. Lubich, Modulated Fourier expansions of highly oscillatory differential equations. *Found. Comput. Math.*, **3**, 327–345, 2003.

[5] D. Cohen, E. Hairer and C. Lubich, Numerical energy conservation for multi-frequency oscillatory differential equations. *BIT*, **45**, 287-305, 2005.

[6] D. Cohen, E. Hairer and C. Lubich, Long-time analysis of nonlinearly perturbed wave equations via modulated Fourier expansions. *Arch. Ration. Mech. Anal.*, **187**, 341–368, 2008.

[7] D. Cohen, E. Hairer and C. Lubich, Conservation of energy, momentum and actions in numerical discretizations of nonlinear wave equations. *Numer. Math.*, **110**, 113–143, 2008.

[8] M. Condon, A. Deaño and A. Iserles, On highly oscillatory problems arising in electronic engineering. *M2AN*, **43**, 785–804, 2009.

[9] M. Condon, A. Deaño and A. Iserles, On second order differential equations with highly oscillatory forcing terms. *Proc. Royal Soc. A*, **466**, 1809–1828, 2010.

[10] P. Console and E. Hairer, Long-term stability of symmetric partitioned linear multistep methods. *Preprint*, 2011.

[11] E. Faou, L. Gauckler and C. Lubich, Sobolev stability of plane wave solutions to the cubic nonlinear Schrödinger equation on a torus. *Preprint*, 2011.

[12] E. Fermi, J. Pasta and S. Ulam, Studies of non linear problems. Technical Report LA-1940, Los Alamos, 1955.

[13] L. Galgani, A. Giorgilli, A. Martinoli and S. Vanzini, On the problem of energy equipartition for large systems of the Fermi-Pasta-Ulam type: analytical and numerical estimates. *Physica D*, **59**, 334–348, 1992.

[14] G. Gallavotti *The Fermi-Pasta-Ulam problem. A status report.* Lecture Notes in Physics, **728**, Springer, Berlin, 2008.

[15] B. García-Archilla, J. M. Sanz-Serna and R. D. Skeel, Long-time-step methods for oscillatory differential equations. *SIAM J. Sci. Comput.*, **20**, 930–963, 1999.

[16] L. Gauckler, *Long-time analysis of Hamiltonian partial differential equations and their discretizations.* Ph. D. thesis, Univ. Tübingen, 2010.

[17] L. Gauckler, E. Hairer, C. Lubich and D. Weiss, Metastable energy strata in weakly nonlinear wave equations. *Preprint*, 2011.

[18] L. Gauckler and C. Lubich, Nonlinear Schrödinger equations and their spectral semi-discretizations over long times. *Found. Comput. Math.*, **10**, 141–169, 2010.

[19] L. Gauckler and C. Lubich, Splitting integrators for nonlinear Schrödinger equations over long times. *Found. Comput. Math.*, **10**, 275–302, 2010.

[20] V. Grimm and M. Hochbruck, Error analysis of exponential integrators for oscillatory second-order differential equations. *J. Phys. A*, **39**(19), 5495–5507, 2006.

[21] E. Hairer and C. Lubich, Long-time energy conservation of numerical methods for oscillatory differential equations. *SIAM J. Numer. Anal.*, **38**, 414–441, 2001.

[22] E. Hairer and C. Lubich, Symmetric multistep methods over long times. *Numer. Math.*, **97**, 699–723, 2004.

[23] E. Hairer and C. Lubich, Spectral semi-discretisations of weakly nonlinear wave equations over long times. *Found. Comput. Math.*, **8**, 319–334, 2008.

[24] E. Hairer and C. Lubich, On the energy distribution in Fermi-Pasta-Ulam lattices. *Preprint*, 2010.

[25] E. Hairer, C. Lubich and G. Wanner, *Geometric Numerical Integration. Structure-Preserving Algorithms for Ordinary Differential Equations.* Springer Series in Computational Mathematics 31, Springer-Verlag, Berlin, 2nd edition, 2006.

[26] G. D. Quinlan and S. Tremaine, Symmetric multistep methods for the numerical integration of planetary orbits. *Astron. J.*, **100**, 1694–1700, 1990.

[27] J. M. Sanz-Serna, Modulated Fourier expansions and heterogeneous multiscale methods. *IMA J. Numer. Anal.*, **29**, 595–605, 2009.

[28] Y.-F. Tang, The symplecticity of multi-step methods. *Computers Math. Applic.*, **25**, 83–90, 1993.

7

The Dual Role of Convection in 3D Navier-Stokes Equations

Thomas Y. Hou
Applied and Comput. Math.
Caltech

Zuoqiang Shi
Mathematical Sciences Center
Tsinghua University

Shu Wang
College of Applied Sciences
Beijing University of Technology

Abstract

We investigate the dual role of convection on the large time behavior of the 3D incompressible Navier-Stokes equations. On the one hand, convection is responsible for generating small scales dynamically. On the other hand, convection may play a stabilizing role in potentially depleting nonlinear vortex stretching for certain flow geometry. Our study is centered around a 3D model that was recently proposed by Hou and Lei in [23] for axisymmetric 3D incompressible Navier-Stokes equations with swirl. This model is derived by neglecting the convection term from the reformulated Navier-Stokes equations and shares many properties with the 3D incompressible Navier-Stokes equations. In this paper, we review some of the recent progress in studying the singularity formation of this 3D model and how convection may destroy the mechanism that leads to singularity formation in the 3D model.

7.1 Introduction

Whether the 3D incompressible Navier-Stokes equations can develop a finite time singularity from smooth initial data with finite energy is one of the seven Millennium problems posted by the Clay Mathematical Institute [16]. This problem is challenging because the vortex stretching nonlinearity is super-critical for the 3D Navier-Stokes equation. Conventional functional analysis based on energy type estimates fails to provide a definite answer to this problem. Global regularity results are obtained only under certain smallness assumptions on the initial data or the solution itself. Due to the incompressibility condition, the convection term seems to be neutrally stable if one tries to estimate the L^p $(1 < p \leq \infty)$ norm of the vorticity field. As a result, the main effort has been to use the diffusion term to control the nonlinear vortex stretching term by diffusion without making use of the convection term explicitly.

In [23], Hou and Lei investigated the role of convection by constructing a new 3D model for axisymmetric 3D incompressible Navier-Stokes equations with swirl. The 3D model is derived based on the reformulated Navier-Stokes equation given below

$$\partial_t u_1 + u^r (u_1)_r + u^z (u_1)_z = \nu\big(\partial_r^2 + \frac{3}{r}\partial_r + \partial_z^2\big)u_1 + 2\partial_z \psi_1 u_1, \tag{7.1}$$

$$\partial_t \omega_1 + u^r (\omega_1)_r + u^z (\omega_1)_z = \nu\big(\partial_r^2 + \frac{3}{r}\partial_r + \partial_z^2\big)\omega_1 + \partial_z\big((u_1)^2\big), \tag{7.2}$$

$$-\big(\partial_r^2 + \frac{3}{r}\partial_r + \partial_z^2\big)\psi_1 = \omega_1, \tag{7.3}$$

where $u_1 = u^\theta/r$, $\omega_1 = \omega^\theta/r$, $\psi_1 = \psi^\theta/r$. Here u^θ, ω^θ ψ^θ are the angular velocity, angular vorticity and angular stream-function, respectively. The radial velocity u^r and the axial velocity u^z are given by $u^r = -r(\psi_1)_z$ and $u^z = (r^2\psi_1)_r/r$. The 3D model of Hou-Lei is obtained by simply dropping the convection term in the reformulated Navier-Stokes equations (7.1)–(7.3), which is given by the following nonlinear nonlocal system

$$\partial_t u_1 = \nu\big(\partial_r^2 + \frac{3}{r}\partial_r + \partial_z^2\big)u_1 + 2\partial_z \psi_1 u_1, \tag{7.4}$$

$$\partial_t \omega_1 = \nu\big(\partial_r^2 + \frac{3}{r}\partial_r + \partial_z^2\big)\omega_1 + \partial_z\big((u_1)^2\big), \tag{7.5}$$

$$-\big(\partial_r^2 + \frac{3}{r}\partial_r + \partial_z^2\big)\psi_1 = \omega_1. \tag{7.6}$$

Note that (7.4)–(7.6) is already a closed system. This model preserves almost all the properties of the full 3D Navier-Stokes equations, including the energy identity for smooth solutions of the 3D model, the non-blowup criterion of Beale-Kato-Majda type [1], the non-blowup criterion of Prodi-Serrin type [34, 35], and the partial regularity result [24] which is an analogue of the well-known Caffarelli-Kohn-Nirenberg theory [2] for the full Navier-Stokes equations.

One of the main findings of [23] is that the 3D model (7.4)–(7.6) has a very different behavior from that of the full Navier-Stokes equations although it shares many properties with those of the Navier-Stokes equations. In [23], the authors presented numerical evidence which supports the notion that the 3D model may develop a potential finite time singularity. However, the Navier-Stokes equations with the same initial data seem to have a completely different behavior.

In a recent paper [26], we rigorously proved the finite time singularity formation of this 3D model for a class of initial boundary value problems with smooth initial data of finite energy. The analysis of the finite time singularity for the 3D model was rather subtle. Currently, there is no systematic method of analysis available to study singularity formation of a nonlinear nonlocal system. In [26], we introduced an effective method of analysis to study singularity formation of this nonlinear nonlocal multi-dimensional system. The initial boundary value problem considered in [26] uses a mixed Dirichlet Robin boundary condition. The local well-posedness of this mixed initial boundary problem is nontrivial. In this paper, we provide a rigorous proof of the local well-posedness of the 3D model with this mixed Dirichlet Robin boundary problem.

We remark that formation of singularities for various model equations for the 3D Euler equations or the surface quasi-geostrophic equation has been investigated by Constantin-Lax-Majda [9], Constantin [5], DeGregorio [12, 13], Cordoba-Cordoba-Fontelos [8], Chae-Cordoba-Cordoba-Fontelos [4], and Li-Rodrigo [30]. In a recent paper related to the present one, Hou, Li, Shi, Wang and Yu [25] have proved the finite time singularity of a one-dimensional nonlinear nonlocal system:

$$u_t = 2uv, \quad v_t = H(u^2),$$

where H is the Hilbert transform. This is a simplified system of the original 3D model along the symmetry axis. Here v plays the same role as ψ_z. The singularity of this nonlocal system is remarkably similar to that of the 3D model.

The work of Hou and Lei [23] was motivated by the recent study of Hou and Li in [22], where the authors studied the stabilizing effect of convection via a new 1D model. They proved dynamic stability of this 1D model by exploiting the stabilizing effect of convection and constructing a Lyapunov function. A surprising result from their study is that there is a beautiful cancellation between the convection term and the nonlinear stretching term when one constructs an appropriate Lyapunov function. This Lyapunov estimate gives rise to a global pointwise estimate for the derivatives of the vorticity in their model.

We would like to emphasize that the study of [22, 23] is based on a reduced model for certain flow geometry. It is premature to conclude that the convection term could lead to depletion of singularity of the Navier-Stokes equations in general. The convection term may act as a destabilizing term for a different flow geometry. A main message from this line of study is that the convection term carries important physical information. We need to take the convection term into consideration in an essential way in our analysis of the Navier-Stokes equations.

The rest of the paper is organized as follows. In Section 2, we discuss the role of convection from the Lagrangian perspective and present some numerical evidence that the local geometric regularity of the vortex lines may deplete the nonlinear vortex stretching dynamically. In Section 3, we investigate the role of convection by studying the potential singular behavior of the 3D model which neglects convection in the reformulated Navier-Stokes equation. We present some theoretical results on finite time singularity formation of the 3D model in Section 4. Finally we present the analysis of the local well-posedness of the 3D model with the mixed Dirichlet Robin boundary condition in Section 5.

7.2 The role of convection from the Lagrangian perspective

Due to the supercritical nature of the nonlinearity of the 3D Navier-Stokes equations, the 3D Navier-Stokes equations with large initial data are convection dominated. Thus the understanding of whether the corresponding 3D Euler equations would develop a finite time blowup could shed useful light on the global regularity of the Navier-Stokes equations.

We consider the 3D Euler equations in the vorticity form. We note that we can rewrite the vorticity equation in a commutator form (or a Lie derivative) as follows:

$$\boldsymbol{\omega}_t + (\mathbf{u} \cdot \nabla)\boldsymbol{\omega} - (\boldsymbol{\omega} \cdot \nabla)\mathbf{u} = 0.$$

Through this commutator formulation, we can see that the convection term may have the potential to dynamically cancel or weaken the vortex stretching term under certain geometric regularity conditions.

Another way to realize the importance of convection is to use the Lagrangian formulation of the vorticity equation. When we consider the two terms together, we preserve the Lagrangian structure of the solution

[32]:

$$\boldsymbol{\omega}(X(\alpha,t),t) = X_\alpha(\alpha,t)\boldsymbol{\omega}_0(\alpha), \tag{7.7}$$

where $X_\alpha = \frac{\partial X}{\partial \alpha}$ and $X(\alpha,t)$ is the flow map:

$$\frac{dX}{dt}(\alpha,t) = \mathbf{u}(X(\alpha,t),t), \quad X(\alpha,0) = \alpha.$$

We believe that (7.7) is an important signature of the 3D incompressible Euler equation. An immediate consequence of (7.7) is that vorticity increases in time only through the dynamic deformation of the Lagrangian flow map, which is volume preserving, i.e. $\det(X_\alpha(\alpha,t)) \equiv 1$. Thus, as vorticity increases dynamically, the parallelepiped spanned by the three vectors, $(X_{\alpha_1}, X_{\alpha_2}, X_{\alpha_3})$, will experience severe deformation and become flattened dynamically. Such deformation tends to weaken the nonlinearity of vortex stretching dynamically.

7.2.1 A brief review

In this subsection, we give a brief review of some of the theoretical and computational studies of the 3D Euler equation. Due to the formal quadratic nonlinearity in vortex stretching, classical solutions of the 3D Euler equation are known to exist only for a short time [32]. One of the most well-known non-blowup results on the 3D Euler equations is due to Beale-Kato-Majda [1] who showed that the solution of the 3D Euler equations blows up at T if and only if $\int_0^T \|\boldsymbol{\omega}\|_\infty(t)\, dt = \infty$, where $\boldsymbol{\omega}$ is vorticity.

There have been some interesting recent theoretical developments. In particular, Constantin-Fefferman-Majda [6] showed that local geometric regularity of the unit vorticity vector can lead to depletion of the vortex stretching. Denote $\boldsymbol{\xi} = \boldsymbol{\omega}/|\boldsymbol{\omega}|$ as the unit vorticity vector and $\mathbf{u}$ the velocity field. Roughly speaking, Constantin-Fefferman-Majda proved that if (1) $\|\mathbf{u}\|_\infty$ is bounded in a $O(1)$ region containing the maximum vorticity, and (2) $\int_0^t \|\nabla\boldsymbol{\xi}\|_\infty^2 d\tau$ is uniformly bounded for $t < T$, then the solution of the 3D Euler equations remains regular up to $t = T$.

There has been considerable effort put into computing a finite time singularity of the 3D Euler equation. The finite time collapse of two anti-parallel vortex tubes by R. Kerr [28, 29] has received a lot of attention. With resolution of order $512 \times 256 \times 192$, his computations showed that the maximum vorticity blows up like $O((T-t)^{-1})$ with $T = 18.9$. In his subsequent paper [29], Kerr applied a high wave number filter to the data obtained in his original computations to "remove the noise that masked the structures in earlier graphics" presented in [28]. The singularity time was revised to $T = 18.7$. Kerr's blowup scenario is consistent with the Beale-Kato-Majda non-blowup criterion [1] and the Constantin-Fefferman-Majda non-blowup criterion [6]. It is worth noting that there is still a considerable gap between the predicted singularity time $T = 18.7$ and the final time $t = 17$ of Kerr's original computations which he used as the primary evidence for the finite time singularity.

7.2.2 The local non-blowup criteria of Deng-Hou-Yu [10, 11]

Motivated by the result of [6], Deng, Hou and Yu [10] have obtained a sharper non-blowup condition which uses a Lagrangian approach and the very localized information of the vortex lines. More specifically, they assume that at each time t there exists some vortex line segment L_t on which the local maximum vorticity is comparable to the global maximum vorticity. Further, they denote $L(t)$ as the arclength of L_t, $\mathbf{n}$ the unit normal vector of L_t, and κ the curvature of L_t. If (1) $\max_{L_t}(|\mathbf{u}\cdot\boldsymbol{\xi}| + |\mathbf{u}\cdot\mathbf{n}|) \leq C_U(T-t)^{-A}$ with $A < 1$, and (2) $C_L(T-t)^B \leq L(t) \leq C_0/\max_{L_t}(|\kappa|, |\nabla\cdot\boldsymbol{\xi}|)$ for $0 \leq t < T$, then they show that the solution of the 3D Euler equations remains regular up to $t = T$ provided that $A + B < 1$.

In Kerr's computations, the first condition of Deng-Hou-Yu's non-blowup criterion is satisfied with $A = 1/2$ if we use $\|\mathbf{u}\|_\infty \leq C(T-t)^{-1/2}$ as alleged in [29]. Kerr's computations suggested that κ and $\nabla\cdot\boldsymbol{\xi}$ are bounded by $O((T-t)^{-1/2})$ in the inner region of size $(T-t)^{1/2} \times (T-t)^{1/2} \times (T-t)$ [29]. Moreover, the length of the vortex tube in the inner region is of order $(T-t)^{1/2}$. If we choose a vortex line segment of length $(T-t)^{1/2}$ (i.e. $B = 1/2$), then the second condition is satisfied. However, this would violate the condition $A+B < 1$. Thus Kerr's computations fall into the critical case of the non-blowup criterion of [10]. In a subsequent paper [11], Deng-Hou-Yu improved the non-blowup condition to include the critical case $A + B = 1$, with some additional constraint on the scaling constants.

We remark that in a recent paper [27], Hou and Shi introduced a different method of analysis to study the non-blowup criterion of the 3D Euler and the SQG model. By performing estimates on the integral of the absolute value of vorticity along a local vortex line segment, they established a relatively sharp dynamic growth estimate of maximum vorticity under some mild assumptions on the local geometric regularity of the vorticity vector. Under some additional assumption on the vorticity field, which seems to be consistent with the computational results of [19], they proved that the maximum vorticity can not grow faster than double exponential in time. This analysis extends to some extent the earlier results by Cordoba-Fefferman [7] and Deng-Hou-Yu [10, 11].

7.2.3 Computing potentially singular solutions using pseudo-spectral methods

It is an extremely challenging task to compute a potential Euler singularity numerically. First of all, it requires a tremendous amount of numerical resolution in order to capture the nearly singular behavior of the Euler equations. Secondly, one must perform a careful convergence study. It is risky to interpret the blowup of an under-resolved computation as evidence of finite time singularities for the 3D Euler equations. Thirdly, we need to validate the asymptotic blowup rate, i.e. is the blowup rate $\|\boldsymbol{\omega}\|_{L^\infty} \approx \frac{C}{(T-t)^\alpha}$ asymptotically valid as $t \to T$? If a numerical solution is well resolved only up to T_0 and there is still an $O(1)$ gap between T_0 and the predicted singularity time T, then one can not apply the Beale-Kato-Majda criterion [1] to this extrapolated singularity since the most significant contribution to $\int_0^T \|\boldsymbol{\omega}(t)\|_{L^\infty} dt$ comes from the time interval $[T_0,\, T]$. But ironically there is no accuracy in the extrapolated solution in this time interval if $(T - T_0) = O(1)$. Finally, the blowup rate of the numerical solution must be consistent with other non-blowup criteria [6, 10, 11]. Guidance from analysis is clearly needed.

In [19], Hou and Li performed high resolution computations of the 3D Euler equations using the two-antiparallel vortex tubes initial data. They used the same initial condition whose analytic formula was given by [28]. They used two different pseudo-spectral methods. The first pseudo-spectral method used the standard 2/3 de-aliasing rule to remove the aliasing error. For the second pseudo-spectral method, they used a novel 36th order Fourier smoothing to remove the aliasing error. In order to perform a careful resolution study, they used a sequence of resolutions: $768 \times 512 \times 1536$, $1024 \times 768 \times 2048$ and $1536 \times 1024 \times 3072$ in their computations. They computed the solution up to $t = 19$, beyond the alleged singularity time $T = 18.7$ by Kerr [29].

We first illustrate the dynamic evolution of the vortex tubes. Figure 7.2 describes the isosurface of the 3D vortex tubes at $t = 0$ and $t = 6$, respectively. As we can see, the two initial vortex tubes are very smooth

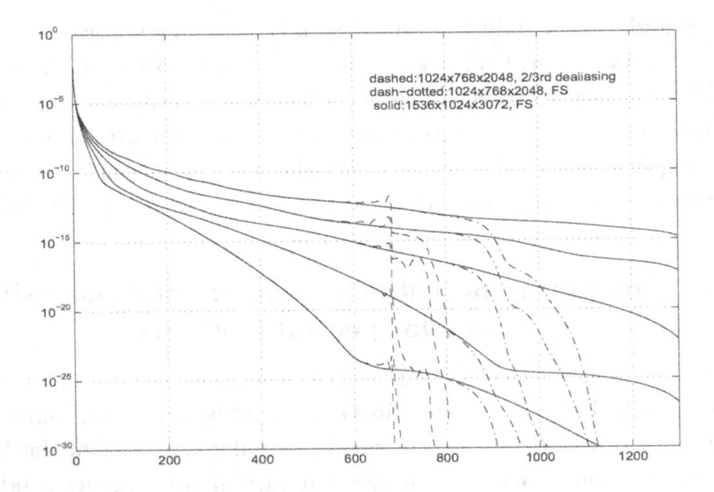

Figure 7.1 The energy spectra versus wave numbers. The dashed lines and dashed-dotted lines are the energy spectra with the resolution $1024 \times 768 \times 2048$ using the 2/3 de-aliasing rule and Fourier smoothing, respectively. The times for the spectra lines are at $t = 15, 16, 17, 18, 19$, respectively.

and relatively symmetric. As time evolves, the two vortex tubes approach each other and become flattened dynamically. By time $t = 6$ there is already a significant flattening near the center of the tubes. In Figure 7.3 we plot the local 3D vortex structure of the upper vortex tube at $t = 17$. By this time the vortex tube has turned into a thin vortex sheet with rapidly decreasing thickness. We observe that the vortex lines become relatively straight and the vortex sheet rolls up near the left edge of the sheet.

We now perform a convergence study for the two numerical methods using a sequence of resolutions. For the Fourier smoothing method, we use the resolutions $768 \times 512 \times 1536$, $1024 \times 768 \times 2048$, and $1536 \times 1024 \times 3072$, respectively, whereas the 2/3 de-aliasing method uses the resolutions $512 \times 384 \times 1024$, $768 \times 512 \times 1536$ and $1024 \times 768 \times 2048$, respectively.

In Figure 7.1 we compare the Fourier spectra of the energy obtained by using the 2/3 de-aliasing method with those obtained by the Fourier smoothing method. For a fixed resolution $1024 \times 768 \times 2048$, the Fourier spectra obtained by the Fourier smoothing method retain more effective Fourier modes than those obtained by the 2/3 de-aliasing method and does not give the spurious oscillations in the Fourier spectra. In

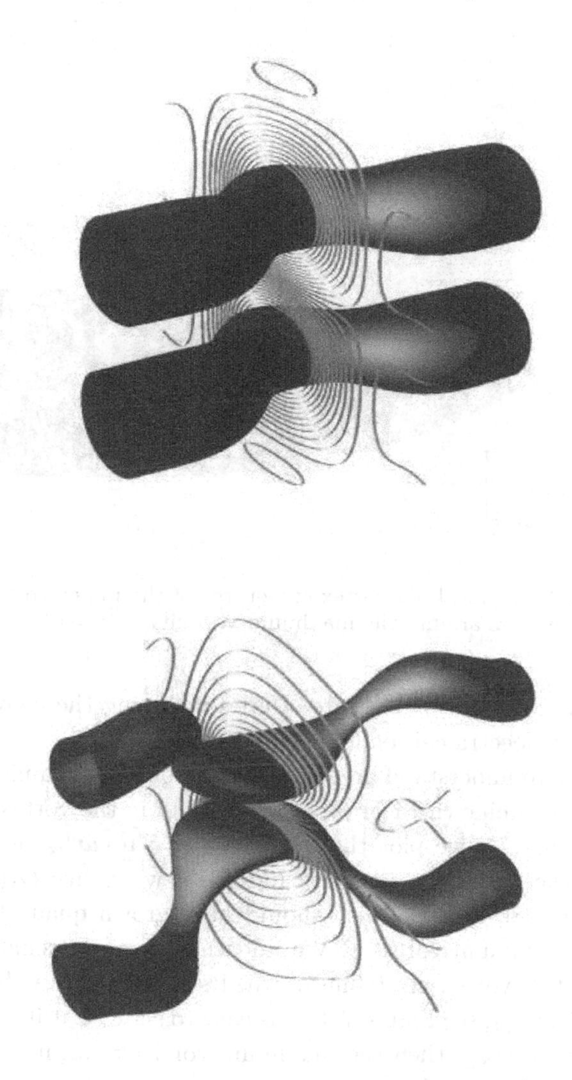

Figure 7.2 The 3D view of the vortex tube for $t = 0$ and $t = 6$. The tube is the isosurface at 60% of the maximum vorticity. The ribbons on the symmetry plane are the contours at other different values.

comparison, the Fourier spectra obtained by the 2/3 de-aliasing method produce some spurious oscillations near the 2/3 cut-off point. It is important to emphasize that the Fourier smoothing method conserves the

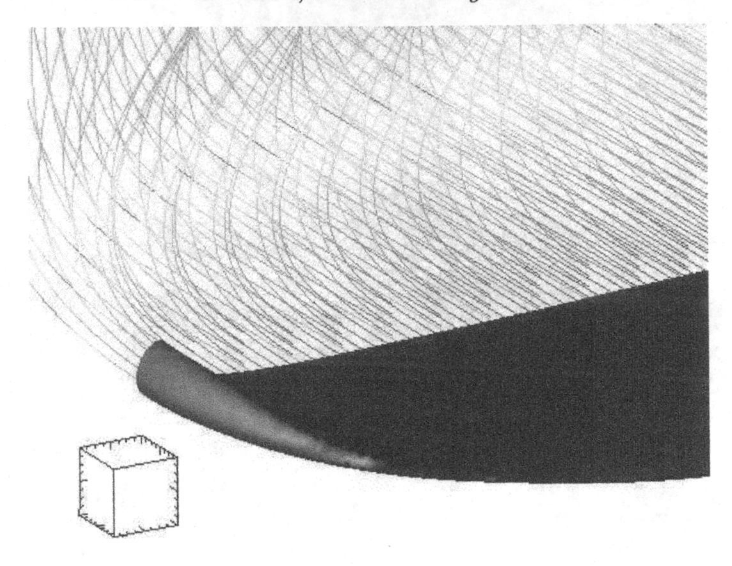

Figure 7.3 The local 3D vortex structures of the upper vortex tube and vortex lines around the maximum vorticity at $t = 17$.

total energy extremely well. More studies including the convergence of the enstrophy spectra can be found in [19, 20, 21].

To gain more understanding of the nature of the dynamic growth in vorticity, we examine the degree of nonlinearity in the vortex stretching term. In Figure 7.4 we plot the quantity $\|\xi \cdot \nabla \mathbf{u} \cdot \boldsymbol{\omega}\|_\infty$ as a function of time. If the maximum vorticity indeed blew up like $O((T-t)^{-1})$, as alleged in [28], this quantity should have grown quadratically as a function of maximum vorticity. We find that there is tremendous cancellation in this vortex stretching term. Its growth rate is bounded by $C\|\vec{\omega}\|_\infty \log(\|\vec{\omega}\|_\infty)$, see Figure 7.4. It is easy to show that if $\|\xi \cdot \nabla \mathbf{u} \cdot \boldsymbol{\omega}\|_\infty \leq C\|\vec{\omega}\|_\infty \log(\|\vec{\omega}\|_\infty)$, then the maximum vorticity can not grow faster than doubly exponential in time.

In the right plot of Figure 7.4, we plot the double logarithm of the maximum vorticity as a function of time. We observe that the maximum vorticity indeed does not grow faster than doubly exponential in time. We have also examined the growth rate of maximum vorticity by extracting the data from Kerr's paper [28]. We find that $\log(\log(\|\boldsymbol{\omega}\|_\infty))$ basically scales linearly with respect to t from $14 \leq t \leq 17.5$ when his computations are still reasonably resolved. This implies that the maximum vorticity up to $t = 17.5$ in Kerr's computations does not grow

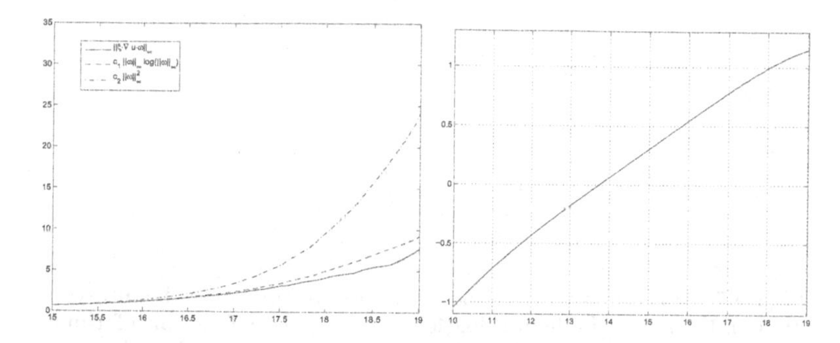

Figure 7.4 Left plot: Study of the vortex stretching term in time, resolution 1536 × 1024 × 3072. The solid line corresponds to $\|\boldsymbol{\xi}\cdot\nabla\mathbf{u}\cdot\boldsymbol{\omega}\|_\infty$, and the dashed and dash-dotted lines correspond to $c_1\|\boldsymbol{\omega}\|_\infty \log(\|\boldsymbol{\omega}\|_\infty)$ and $c_2\|\boldsymbol{\omega}\|_\infty^2$ respectively. The fact $|\boldsymbol{\xi}\cdot\nabla\mathbf{u}\cdot\boldsymbol{\omega}| \leq c_1|\boldsymbol{\omega}|\log|\boldsymbol{\omega}|$ plus $\frac{D}{Dt}|\boldsymbol{\omega}| = \boldsymbol{\xi}\cdot\nabla\mathbf{u}\cdot\boldsymbol{\omega}$ implies $|\boldsymbol{\omega}|$ bounded by doubly exponential. Right plot: $\log\log\|\boldsymbol{\omega}\|_\infty$ vs time.

faster than doubly exponential in time, which is consistent with our conclusion.

7.3 Numerical evidence of finite time singularity of the 3D model

As we mentioned in the Introduction, the 3D model shares many properties with the full 3D Navier-Stokes equations at the theoretical level. In this section, we will demonstrate that the 3D model without the convection term has a very different behavior from the full Navier-Stokes equation. In particular, we present numerical evidence based on the computations of [23] that seems to suggest that the 3D model develops a potential finite time singularity from smooth initial data with finite energy. However, the mechanism for developing a finite time singularity of the 3D model seems to be destroyed when we add the convection term back to the 3D model. This illustrates the important role played by convection from a different perspective.

By exploiting the axisymmetric geometry of the problem, Hou and Lei obtained a very efficient adaptive solver with effective local resolutions of order 4096^3. More specifically, since the potential singularity must appear along the symmetry axis at $r = 0$, they used the following coordinate transformation along the r-direction to achieve the adaptivity by

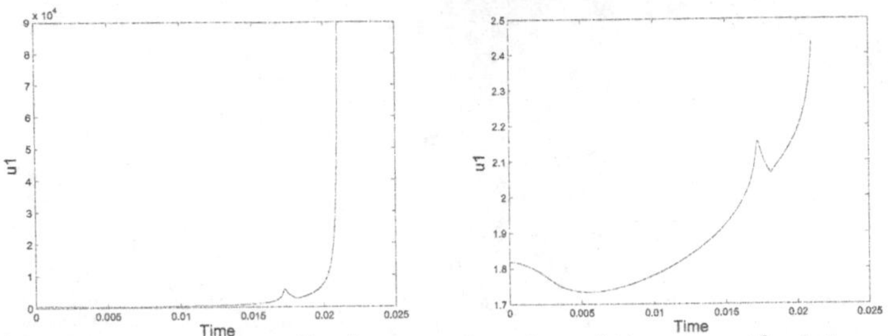

Figure 7.5 Left figure: $\|u_1\|_\infty$ as a function of time over the interval $[0,\ 0.021]$. The right figure: $\log(\log(\|u_1\|_\infty))$ as a function of time over the same interval. The solution is computed by the adaptive mesh with $N_z = 4096$, $N_r = 400$, $\Delta t = 2.5 \times 10^{-7}$, $\nu = 0.001$.

clustering the grid points near $r = 0$:

$$r = f(\alpha) \equiv \alpha - 0.9 \sin(\pi\alpha)/\pi.$$

With this level of resolution, they obtained an excellent fit for the asymptotic blowup rate of maximum axial vorticity.

The initial condition we consider in our numerical computations is given by

$$u_1(z, r, 0) = (1 + \sin(4\pi z))(r^2 - 1)^{20}(r^2 - 1.2)^{30},$$
$$\psi_1(z, r, 0) = 0,$$
$$\omega_1(z, r, 0) = 0.$$

A second order finite difference discretization is used in space, and the classical fourth order Runge-Kutta method is used to discretize in time.

In the following, we present numerical evidence which seems to support the notion that u_1 may develop a potential finite time singularity for the initial condition we consider. In Figure 7.5 we plot the maximum of u_1 in time over the time interval $[0,\ 0.021]$ using the adaptive mesh method with $N_z = 4096$ and $N_r = 400$. The time step is chosen to be $\Delta t = 2.5 \times 10^{-7}$. We observe that $\|u_1\|_\infty$ experiences a very rapid growth in time after $t = 0.02$. In Figure 7.5 (the right plot), we also plot $\log(\log(\|u_1\|_\infty))$ as a function of time. It is clear that $\|u_1\|_\infty$ grows much faster than double exponential in time.

To obtain further evidence for a potential finite time singularity, we study the asymptotic growth rate of $\|u_1\|_\infty$ in time. We look for a finite time singularity of the form:

$$\|u_1\|_\infty \approx \frac{C}{(T - t)^\alpha}.$$

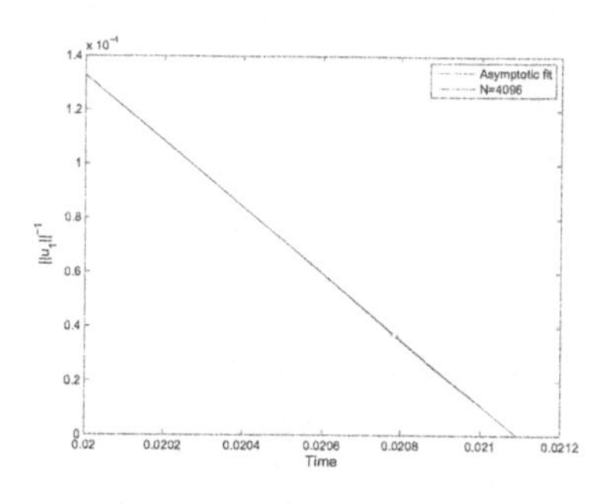

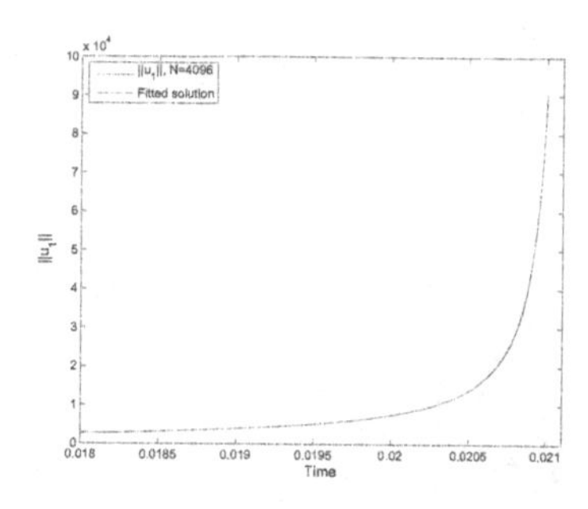

Figure 7.6 The left plot: The inverse of $||u_1||_\infty$ (dark) versus the asymptotic fit (gray) for the viscous model. The right plot: $||u_1||_\infty$ (dark) versus the asymptotic fit (gray). The asymptotic fit is of the form: $||u_1||_\infty^{-1} \approx \frac{(T-t)}{C}$ with $T = 0.02109$ and $C = 8.20348$. The solution is computed by an adaptive mesh with $N_z = 4096$, $N_r = 400$, $\Delta t = 2.5 \times 10^{-7}$, $\nu = 0.001$. In each case, the two curves are indistinguishable at normal resolution.

We find that the inverse of $||u_1||_\infty$ is almost a perfect linear function of time, see Figure 7.6. By using a least square fit of the inverse of $||u_1||_\infty$, we find the best fit for α, the potential singularity time T and the constant C. In Figure 7.6 (the left plot), we plot $||u_1||_\infty^{-1}$ as a function of time. We can see that the agreement between the computed solution with $N_z \times N_r = 4096 \times 400$ and the fitted solution is almost perfect. In the right box of Figure 7.6, we plot $||u_1||_\infty$ computed by our adaptive method against the form fit $C/(T-t)$ with $T = 0.02109$ and $C = 8.20348$. The two curves are almost indistinguishable during the final stage of the computation from $t = 0.018$ to $t = 0.021$. Note that u_1 has the same scaling as the axial vorticity. Thus, the $O(1/(T-t))$ blowup rate of u_1 is consistent with the non-blowup criterion of Beale-Kato-Majda type.

We present the 3D view of u_1 as a function of r and z in Figures 7.7 and 7.8. We note that u_1 is symmetric with respect to $z = 0.375$ and w_1 is anti-symmetric with respect to $z = 0.375$. The support of the solution u_1 in the most singular region is isotropic and appears to be locally self-similar.

To study the dynamic alignment of the vortex stretching term, we plot the solution u_1 on top of $\psi_{1,z}$ along the symmetry axis $r = 0$ at t=0.021 in Figure 7.9. We observe that there is a significant overlap between the support of the maximum of u_1 and that of the maximum of $\psi_{1,z}$. Moreover, the solution u_1 has a strong alignment with $\psi_{1,z}$ near the

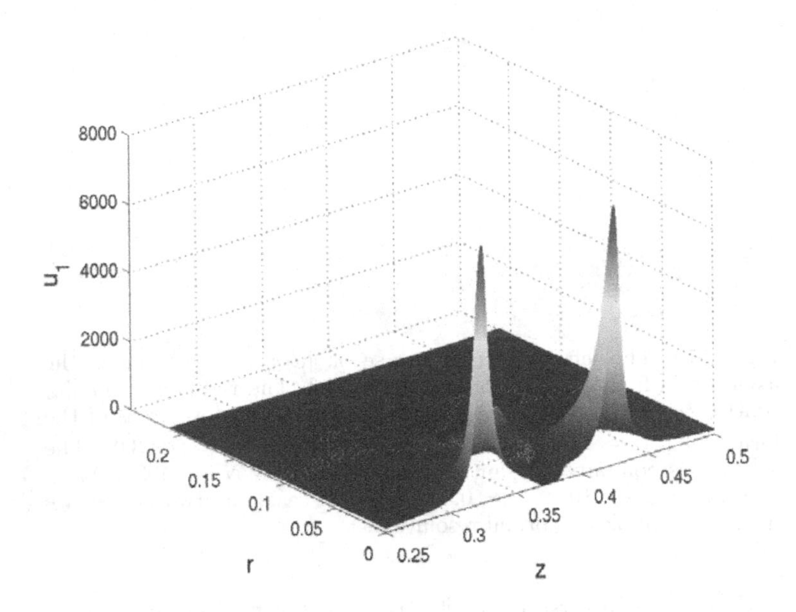

Figure 7.7 The 3D view of u_1 at $t = 0.02$ for the viscous model computed by the adaptive mesh with $N_z = 4096$, $N_r = 400$, $\Delta t = 2.5 \times 10^{-7}$, $\nu = 0.001$.

region of the maximum of u_1. The local alignment between u_1 and $\psi_{1,z}$ induces a strong nonlinearity on the right hand side of the u_1 equation. This strong alignment between u_1 and $\psi_{1,z}$ is the main mechanism for the potential finite time blowup of the 3D model.

It is interesting to see how convection may change the dynamic alignment of the vortex stretching term in the 3D model. We add the convection term back to the 3D model and use the solution of the 3D model at $t = 0.02$ as the initial condition for the full Navier-Stokes equations. We observe that the local alignment between u_1 and $\psi_{1,z}$ is destroyed for the full Navier-Stokes equations. As a result, the solution becomes defocused and smoother along the symmetry axis, see Figure 7.10. As time evolves, the two focusing centers approach each other. This process creates a strong internal layer orthogonal to the z-axis. The solution forms a jet that moves away from the symmetry axis (the z-axis) and generates many interesting vortex structures. By the Caffarelli-Kohn-Nirenberg theory, the singularity of the 3D axisymmetric Navier-Stokes equations must be along the symmetry axis. The fact that the most

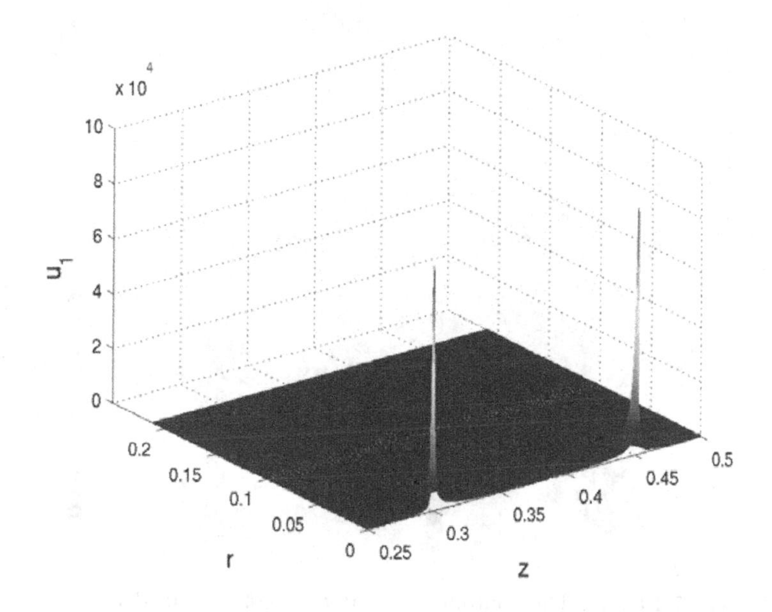

Figure 7.8 The 3D view of u_1 at $t = 0.021$ for the viscous model computed by the adaptive mesh with $N_z = 4096$, $N_r = 400$, $\Delta t = 2.5 \times 10^{-7}$, $\nu = 0.001$.

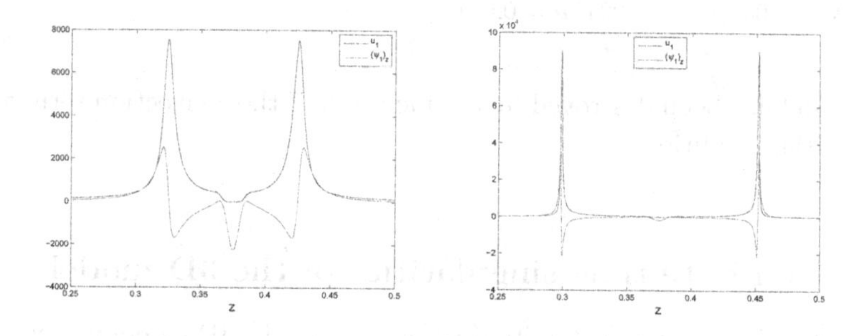

Figure 7.9 u_1 (dark) versus $\psi_{1,z}$ (gray) of the viscous model along the symmetry axis $r = 0$. The left figure corresponds to $t = 0.02$. The right figure corresponds to $t = 0.021$. Adaptive mesh computation with $N_z = 4096$, $N_r = 400$, $\Delta t = 2.5 \times 10^{-7}$, $\nu = 0.001$.

singular part of the solution moves away from the symmetry axis suggests that the mechanism for generating the finite time singularity of the

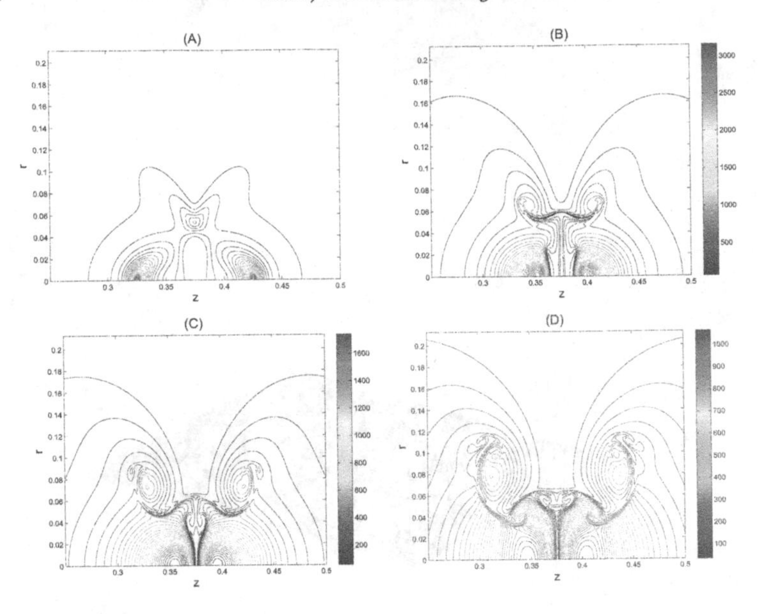

Figure 7.10 (A): The contour of u_1 at $t = 0.02$ obtained from the 3D viscous model which serves as the initial condition for the full Navier-Stokes equations. (B): The contour of u_1 at $t = 0.021$ obtained by solving the full Navier-Stokes equations. (C) and (D): the contours of u_1 at $t = 0.022$ and $t = 0.0235$, respectively, by solving the full Navier-Stokes equations. Adaptive mesh computation with $N_z = 2048$, $N_r = 1024$, $\nu = 0.001$.

3D model has been destroyed by the inclusion of the convection term for this initial condition.

7.4 Finite time singularities of the 3D model

The numerical evidence of finite time blow-up of the 3D model motivates us to prove finite time singularities of the 3D model rigorously. In a recent paper [26], we developed a new method of analysis and proved rigorously that the 3D model develops finite time singularities for a class of initial boundary value problems with smooth initial data of finite energy. In our analysis, we considered the initial boundary value problem of the generalized 3D model which has the following form (we drop the subscript 1 and substitute (7.6) into (7.5)):

$$u_t = 2u\psi_z, \tag{7.8}$$

$$-\Delta\psi_t = \left(u^2\right)_z, \tag{7.9}$$

where Δ is an n-dimensional Laplace operator with $(\mathbf{x}, z) \equiv (x_1, x_2, ..., x_{n-1}, z)$. Our results apply to any dimension greater than or equal to two ($n \geq 2$). Here we only present our results for $n = 3$. We consider the generalized 3D model in both a bounded domain and in a semi-infinite domain with a mixed Dirichlet Robin boundary condition.

7.4.1 Summary of the main result

In [26], we proved rigorously the following finite time blow-up result for the 3D inviscid model.

Theorem 7.1 *Let $\Omega_{\mathbf{x}} = (0, a) \times (0, a)$, $\Omega = \Omega_{\mathbf{x}} \times (0, b)$ and $\Gamma = \{(\mathbf{x}, z) \mid \mathbf{x} \in \Omega_{\mathbf{x}},\ z = 0\}$. Assume that the initial conditions u_0 and ψ_0 satisfy $u_0 > 0$ for $(\mathbf{x}, z) \in \Omega$, $u_0|_{\partial\Omega} = 0$, $u_0 \in H^2(\Omega)$, $\psi_0 \in H^3(\Omega)$ and ψ satisfies (7.10). Moreover, we assume that ψ satisfies the following mixed Dirichlet Robin boundary conditions:*

$$\psi|_{\partial\Omega\setminus\Gamma} = 0, \quad (\psi_z + \beta\psi)|_\Gamma = 0, \tag{7.10}$$

with $\beta > \frac{\sqrt{2}\pi}{a}\left(\frac{1+e^{-2\pi b/a}}{1-e^{-2\pi b/a}}\right)$. Define

$$\phi(x_1, x_2, z) = \left(\frac{e^{-\alpha(z-b)} + e^{\alpha(z-b)}}{2}\right) \sin\left(\frac{\pi x_1}{a}\right) \sin\left(\frac{\pi x_2}{a}\right)$$

where α satisfies $0 < \alpha < \sqrt{2}\pi/a$ and $2\left(\frac{\pi}{a}\right)^2 \frac{e^{\alpha b} - e^{-\alpha b}}{\alpha(e^{\alpha b} + e^{-\alpha b})} = \beta$. If u_0 and ψ_0 satisfy the following condition:

$$\int_\Omega (\log u_0)\phi d\mathbf{x}dz > 0, \quad \int_\Omega \psi_{0z}\phi d\mathbf{x}dz > 0,$$

then the solution of the 3D inviscid model (7.8) − −(7.9) will develop a finite time singularity in the H^2 norm.

7.4.2 Outline of the singularity analysis

We prove the finite time singularity result of the 3D model by contradiction. The analysis uses the local well-posedness result of the 3D model with the above mixed Dirichlet Robin boundary condition, which will be established in Section 5. By the local well-posedness result, we know that there exists a finite time $T > 0$ such that the initial boundary value problem (7.8)–(7.9) with boundary condition given in the above theorem has a unique smooth solution with $u \in C^1([0,T), H^2(\Omega))$ and $\psi \in C^1([0,T), H^3(\Omega))$. Let T_b be the largest time such that the system (7.8)–(7.9) with initial condition u_0, ψ_0 has a smooth solution with $u \in C^1([0,T_b); H^2(\Omega))$ and $\psi \in C^1([0,T_b); H^3(\Omega))$. We claim that $T_b < \infty$. We prove this by contradiction.

Suppose that $T_b = \infty$. This means that for the given initial data u_0, ψ_0, the system (7.8)–(7.9) has a globally smooth solution $u \in C^1([0,\infty); H^2(\Omega))$ and $\psi \in C^1([0,\infty); H^3(\Omega))$. Note that $u|_{\partial\Omega} = 0$ as long as the solution remains smooth.

There are several important ingredients in our analysis. The first one is that we reformulate the u-equation and use $\log(u)$ as the new variable. With this reformulation, the right hand side of the reformulated u-equation becomes linear. Such reformulation is possible since $u_0 > 0$ in Ω implies that $u > 0$ in Ω as long as the solution remains smooth. We now work with the reformulated system given below:

$$(\log(u))_t = 2\psi_z, \quad (\mathbf{x}, z) \in \Omega, \tag{7.11}$$

$$-\Delta\psi_t = \left(u^2\right)_z. \tag{7.12}$$

The second ingredient is to find an appropriate test function ϕ and work with the weak formulation of (7.11)–(7.12). This test function ϕ is chosen as a positive and smooth eigenfunction in Ω that satisfies the following two conditions simultaneously:

$$-\Delta\phi = \lambda_1\phi, \quad \partial_z^2\phi = \lambda_2\phi, \quad \text{for some } \lambda_1, \lambda_2 > 0, \quad (\mathbf{x}, z) \in \Omega. \tag{7.13}$$

Now we multiply ϕ to (7.11) and ϕ_z to (7.12) and integrate over Ω. Upon performing integration by parts, we obtain by using (7.13) that

$$\frac{d}{dt}\int_\Omega (\log u)\phi d\mathbf{x}dz = 2\int_\Omega \psi_z\phi d\mathbf{x}dz, \tag{7.14}$$

$$\lambda_1\frac{d}{dt}\int_\Omega \psi_z\phi d\mathbf{x}dz = \lambda_2\int_\Omega u^2\phi d\mathbf{x}dz. \tag{7.15}$$

It is interesting to note that all the boundary terms resulting from integration by parts vanish due to the boundary condition of ψ, the property of our eigenfunction ϕ, the specific choice of α defined in Theorem 4.1. We have also used the fact that $u|_{z=0} = u|_{z=b} = 0$. Combining (7.15) with (7.14), we obtain our crucial blow-up estimate:

$$\frac{d^2}{dt^2}\int_\Omega (\log u)\phi d\mathbf{x}dz = \frac{2\lambda_2}{\lambda_1}\int_\Omega u^2\phi d\mathbf{x}dz. \tag{7.16}$$

Further, we note that

$$\int_\Omega \log(u)\phi d\mathbf{x}dz \le \int_\Omega (\log(u))^+\phi d\mathbf{x}dz \le \int_\Omega u\phi d\mathbf{x}dz$$
$$\le \left(\int_\Omega \phi d\mathbf{x}dz\right)^{1/2}\left(\int_\Omega \phi u^2 d\mathbf{x}dz\right)^{1/2} \equiv \frac{2a}{\pi\sqrt{\alpha}}\left(\int_\Omega \phi u^2 d\mathbf{x}dz\right)^{1/2}. \quad (7.17)$$

From (7.16) and (7.17), we establish a sharp nonlinear dynamic estimate for $(\int_\Omega \phi u^2 d\mathbf{x}dz)^{1/2}$, which enables us to prove finite time blowup of the 3D model.

This method of analysis is quite robust and captures very well the nonlinear interaction of the multi-dimensional nonlocal system. As a result, it provides a very effective method to analyze the finite time blowup of the 3D model and gives a relatively sharp blowup condition on the initial and boundary values for the 3D model.

7.4.3 Finite time blow-up of the 3D model with conservative BCs

We can also prove finite time blow-up of the 3D model with a conservative boundary condition in a bounded domain. Specifically, we consider the following initial boundary value problem:

$$\begin{cases} u_t &= 2u\psi_z \\ -\Delta\psi_t &= \left(u^2\right)_z \end{cases}, \quad (\mathbf{x}, z) \in \Omega = \Omega_{\mathbf{x}} \times (0, b), \qquad (7.18)$$

$$\psi|_{\partial\Omega\setminus\Gamma} = 0, \quad \psi_z|_\Gamma = 0, \qquad (7.19)$$

$$\psi|_{t=0} = \psi_0(\mathbf{x}, z), \quad u|_{t=0} = u_0(\mathbf{x}, z) \ge 0,$$

where $\mathbf{x} = (x_1, x_2)$, $\Omega_{\mathbf{x}} = (0, a) \times (0, a)$, $\Gamma = \{(\mathbf{x}, z) \in \Omega \mid \mathbf{x} \in \Omega_{\mathbf{x}},\ z = 0$ or $z = b\}$.

The main result is stated in the following theorem.

Theorem 7.2 *Assume that the initial conditions u_0 and ψ_0 satisfy $u_0 \in H^2(\Omega)$, $u_0|_{\partial\Omega} = 0$, $u_0|_\Omega > 0$, $\psi_0 \in H^3(\Omega)$, and ψ satisfies (7.19). Let*

$$\phi(\mathbf{x}, z) = \frac{e^{-\alpha(z-b)} - e^{\alpha(z-b)}}{2} \sin\frac{\pi x_1}{a} \sin\frac{\pi x_2}{a}, \quad (\mathbf{x}, z) \in \Omega,$$

with $\alpha = \frac{\pi}{a}$, *and*

$$A = \int_{\Omega} (\log u_0)\phi d\mathbf{x}dz, \quad B = 2\int_{\Omega} \psi_{0z}\phi d\mathbf{x}dz,$$

$$r(t) = \frac{2\left(\frac{\pi}{a}\right)^2\left(e^{\alpha b} - e^{-\alpha b}\right)}{2\left(\frac{\pi}{a}\right)^2 - \alpha^2} \int_{\Omega_{\mathbf{x}}} (\psi - \psi_0)|_{z=0} \sin\frac{\pi x_1}{a} \sin\frac{\pi x_2}{a} d\mathbf{x} \le \frac{B}{2}.$$

If $A > 0$, $B > 0$ *and* $r(t) \le \frac{B}{2}$ *as long as* u, ψ *remain regular, then the solution of* (7.18)–(7.19) *will develop a finite time singularity in the* H^2 *norm.*

7.4.4 Global regularity of the 3D inviscid model with small data

In this subsection we study the global regularity of the 3D inviscid model for a class of initial data with some appropriate boundary condition. To simplify the presentation of our analysis, we use u^2 and ψ_z as our new variables. We will define $v = \psi_z$ and still use u to stand for u^2. Then the 3D model now has the form:

$$\begin{cases} u_t &= 4uv \\ -\Delta v_t &= u_{zz} \end{cases}, \quad (\mathbf{x}, z) \in \Omega = (0,\delta)\times(0,\delta)\times(0,\delta). \tag{7.20}$$

We choose the following boundary condition for v:

$$v|_{\partial\Omega} = -4, \tag{7.21}$$

and denote $v|_{t=0} = v_0(\mathbf{x}, z)$ and $u|_{t=0} = u_0(\mathbf{x}, z) \ge 0$.

We prove the following global regularity result for the 3D inviscid model with a family of initial boundary value problems.

Theorem 7.3 *Assume that* $u_0, v_0 \in H^s(\Omega)$ *with* $s \ge 4$, $u_0|_{\partial\Omega} = 0$, $v_0|_{\partial\Omega} = -4$ *and* $v_0 \le -4$ *over* Ω. *Then the solution of* (7.20)–(7.21) *remains regular in* $H^s(\Omega)$ *for all time as long as the following holds*

$$\delta(4C_s + 1)\left(\|v_0\|_{H^s} + C_s\|u_0\|_{H^s}\right) < 1,$$

where C_s *is an interpolation constant. Moreover, we have* $\|u\|_{L^\infty} \le \|u_0\|_{L^\infty}e^{-7t}$, $\|u\|_{H^s(\Omega)} \le \|u_0\|_{H^s(\Omega)}e^{-7t}$ *and* $\|v\|_{H^s(\Omega)} \le C$ *for some constant* C *which depends on* u_0, v_0 *and* s *only.*

7.4.5 Blow-up of the 3D model with partial viscosity

In the previous subsections we considered only the inviscid model. In this subsection we show that the 3D model with partial viscosity can also develop finite time singularities. Specifically, we consider the following initial boundary value problem in a semi-infinite domain:

$$\left\{\begin{array}{rcl} u_t &=& 2u\psi_z \\ \omega_t &=& \left(u^2\right)_z + \nu\Delta\omega \quad , \quad (\mathbf{x}, z) \in \Omega = \Omega_{\mathbf{x}} \times (0, \infty), \\ -\Delta\psi &=& \omega. \end{array}\right. \tag{7.22}$$

The initial and boundary conditions are given as follows:

$$\psi|_{\partial\Omega\setminus\Gamma} = 0, \quad (\psi_z + \beta\psi)\,|_\Gamma = 0, \tag{7.23}$$

$$\omega|_{\partial\Omega\setminus\Gamma} = 0, \quad (\omega_z + \gamma\omega)\,|_\Gamma = 0, \tag{7.24}$$

$$\omega|_{t=0} = \omega_0(\mathbf{x}, z), \quad u|_{t=0} = u_0(\mathbf{x}, z) \geq 0, \tag{7.25}$$

where $\Gamma = \{(\mathbf{x}, z) \in \Omega \mid \mathbf{x} \in \Omega_{\mathbf{x}},\ z = 0\}$.

Now we state the main result of this subsection.

Theorem 7.4 *Assume that* $u_0|_{\partial\Omega} = 0$, $u_{0z}|_{\partial\Omega} = 0$, $u_0|_\Omega > 0$, $u_0 \in H^2(\Omega)$, $\psi_0 \in H^3(\Omega)$, $\omega_0 \in H^1(\Omega)$, ψ_0 *satisfies* (7.23) *and* ω_0 *satisfies* (7.24). *Further, we assume that* $\beta \in S_\infty$ *as defined in Lemma 7.6 and* $\beta > \frac{\sqrt{2}\pi}{a}$, $\gamma = \frac{2\pi^2}{\beta a^2}$. *Let*

$$\phi(\mathbf{x}, z) = e^{-\alpha z} \sin\frac{\pi x_1}{a} \sin\frac{\pi x_2}{a}, \quad (\mathbf{x}, z) \in \Omega,$$

where $\alpha = \frac{2\pi^2}{\beta a^2}$ *satisfies* $0 < \alpha < \sqrt{2}\pi/a$. *Define*

$$A = \int_\Omega (\log u_0)\phi \mathrm{d}\mathbf{x}\mathrm{d}z, \quad B = -\int_\Omega \omega_0\phi_z \mathrm{d}\mathbf{x}\mathrm{d}z, \quad D = \frac{2}{2\left(\frac{\pi}{a}\right)^2 - \alpha^2},$$

$$I_\infty = \int_0^\infty \frac{\mathrm{d}x}{\sqrt{x^3+1}}, \quad T^* = \left(\frac{\pi\alpha^3 D^2 B}{12a}\right)^{-1/3} I_\infty.$$

If $A > 0$, $B > 0$, *and* $T^* < (\log 2)\left(\nu\left(\frac{2\pi^2}{a^2} - \alpha^2\right)\right)^{-1}$, *then the solution of model* (7.22) *with initial and boundary conditions* (7.23)–(7.25) *will develop a finite time singularity before* T^*.

7.5 Local well-posedness of the 3D model with mixed Dirichlet Robin Boundary conditions

In this section we prove the local well-posedness of the 3D model with the mixed Dirichlet Robin boundary conditions considered in the previous section. The 3D model with partial viscosity has the following form:

$$\left\{\begin{array}{rcl} u_t & = & 2u\psi_z \\ \omega_t & = & (u^2)_z + \nu\Delta\omega \\ -\Delta\psi & = & \omega \end{array}\right. , \quad (\mathbf{x}, z) \in \Omega = \Omega_{\mathbf{x}} \times (0, \infty), \quad (7.26)$$

where $\Omega_{\mathbf{x}} = (0, a) \times (0, a)$. Let $\Gamma = \{(\mathbf{x}, z) \mid \mathbf{x} \in \Omega_{\mathbf{x}},\ z = 0\}$. The initial and boundary conditions for (7.26) are given as following:

$$\omega|_{\partial\Omega\setminus\Gamma} = 0, \quad (\omega_z + \gamma\omega)\,|_\Gamma = 0, \quad (7.27)$$

$$\psi|_{\partial\Omega\setminus\Gamma} = 0, \quad (\psi_z + \beta\psi)\,|_\Gamma = 0, \quad (7.28)$$

$$\omega|_{t=0} = \omega_0(\mathbf{x}, z), \quad u|_{t=0} = u_0(\mathbf{x}, z). \quad (7.29)$$

The analysis of finite time singularity formation of the 3D model uses the local well-posedness result of the 3D model. The local well-posedness of the 3D model can be proved by using a standard energy estimate and a mollifier if there is no boundary or if the boundary condition is a standard one, see e.g. [32]. For the mixed Dirichlet Robin boundary condition we consider here, the analysis is a bit more complicated since the mixed Dirichlet Robin condition gives rise to a growing eigenmode.

There are two key ingredients in our local well-posedness analysis. The first one is to design a Picard iteration for the 3D model. The second one is to show that the mapping that generates the Picard iteration is a contraction mapping and the Picard iteration converges to a fixed point of the Picard mapping by using the contraction mapping theorem. To establish the contraction property of the Picard mapping, we need to use the well-posedness property of the heat equation with the same Dirichlet Robin boundary condition as ω. The well-posedness analysis of the heat equation with a mixed Dirichlet Robin boundary has been studied in the literature. The case of $\gamma > 0$ is more subtle because there is a growing eigenmode. Nonetheless, we prove that all the essential regularity properties of the heat equation are still valid for the mixed Dirichlet Robin boundary condition with $\gamma > 0$.

The local existence result of our 3D model with partial viscosity is stated in the following theorem.

Theorem 7.5 *Assume that* $u_0 \in H^{s+1}(\Omega)$, $\omega_0 \in H^s(\Omega)$ *for some*

$s > 3/2$, $u_0|_{\partial\Omega} = u_{0z}|_{\partial\Omega} = 0$ and ω_0 satisfies (7.27). Moreover, we assume that $\beta \in S_\infty$ (or S_b) as defined in Lemma 7.6. Then there exists a finite time $T = T\left(\|u_0\|_{H^{s+1}(\Omega)}, \|\omega_0\|_{H^s(\Omega)}\right) > 0$ such that the system (7.26) with boundary condition (7.27),(7.28) and initial data (7.29) has a unique solution, $u \in C([0,T], H^{s+1}(\Omega))$, $\omega \in C([0,T], H^s(\Omega))$ and $\psi \in C([0,T], H^{s+2}(\Omega))$.

The local well-posedness analysis relies on the following local well-posedness of the heat equation and the elliptic equation with mixed Dirichlet Robin boundary conditions. First, the local well-posedness of the elliptic equation with the mixed Dirichlet Robin boundary condition is given by the following lemma [26]:

Lemma 7.6 *There exists a unique solution $v \in H^s(\Omega)$ to the boundary value problem:*

$$-\Delta v = f, \quad (\mathbf{x}, z) \in \Omega, \tag{7.30}$$

$$v|_{\partial\Omega\backslash\Gamma} = 0, \quad (v_z + \beta v)|_\Gamma = 0, \tag{7.31}$$

if $\beta \in S_\infty \equiv \{\beta \mid \beta \neq \frac{\pi|k|}{a}$ for all $k \in \mathbb{Z}^2\}$, $f \in H^{s-2}(\Omega)$ with $s \geq 2$ and $f|_{\partial\Omega\backslash\Gamma} = 0$. Moreover we have

$$\|v\|_{H^s(\Omega)} \leq C_s\|f\|_{H^{s-2}(\Omega)},$$

where C_s is a constant depending on s, $|k| = \sqrt{k_1^2 + k_2^2}$.

Definition 7.7 Let $\mathcal{K} : H^{s-2}(\Omega) \to H^s(\Omega)$ be a linear operator defined as follows: for all $f \in H^{s-2}(\Omega)$, $\mathcal{K}(f)$ is the solution of the boundary value problem (7.30)–(7.31).

It follows from Lemma 7.6 that for any $f \in H^{s-2}(\Omega)$, we have

$$\|\mathcal{K}(f)\|_{H^s(\Omega)} \leq C_s\|f\|_{H^{s-2}(\Omega)}. \tag{7.32}$$

For the heat equation with the mixed Dirichlet Robin boundary condition, we have the following result.

Lemma 7.8 *There exists a unique solution $\omega \in C([0,T]; H^s(\Omega))$ to the initial boundary value problem:*

$$\omega_t = \nu\Delta\omega, \quad (\mathbf{x}, z) \in \Omega, \tag{7.33}$$

$$\omega|_{\partial\Omega\backslash\Gamma} = 0, \quad (\omega_z + \gamma\omega)|_\Gamma = 0, \tag{7.34}$$

$$\omega|_{t=0} = \omega_0(\mathbf{x}, z), \tag{7.35}$$

for $\omega_0 \in H^s(\Omega)$ *with* $s > 3/2$. *Moreover we have the following estimates in the case* $\gamma > 0$

$$\|\omega(t)\|_{H^s(\Omega)} \leq C(\gamma, s) e^{\nu\gamma^2 t} \|\omega_0\|_{H^s(\Omega)}, \quad t \geq 0, \tag{7.36}$$

and

$$\|\omega(t)\|_{H^s(\Omega)} \leq C(\gamma, s, t) \|\omega_0\|_{L^2(\Omega)}, \quad t > 0. \tag{7.37}$$

Remark 7.9 We remark that the growth factor $e^{\nu\gamma^2 t}$ in (7.36) is absent in the case of $\gamma \leq 0$ since there is no growing eigenmode in this case.

Proof First, we prove the solution of the system (7.33)–(7.35) is unique. Let $\omega_1,\ \omega_2 \in H^s(\Omega)$ be two smooth solutions of the heat equation for $0 \leq t < T$ satisfying the same initial condition and the Dirichlet Robin boundary condition. Let $\omega = \omega_1 - \omega_2$. We will prove that $\omega = 0$ by using an energy estimate and the Dirichlet Robin boundary condition at Γ:

$$\begin{aligned}
\frac{1}{2}\frac{d}{dt}\int_\Omega \omega^2 d\mathbf{x}dz &= \nu \int_\Omega \omega\Delta\omega d\mathbf{x}dz \\
&= -\nu \int_\Omega |\nabla\omega|^2 d\mathbf{x}dz - \nu \int_\Gamma \omega\omega_z d\mathbf{x} \\
&= -\nu \int_\Omega |\nabla\omega|^2 d\mathbf{x}dz + \nu\gamma \int_\Gamma \omega^2 d\mathbf{x} \\
&= -\nu \int_\Omega |\nabla\omega|^2 d\mathbf{x}dz - \nu\gamma \int_\Gamma \int_z^\infty \left(\omega^2\right)_z dz d\mathbf{x} \\
&= -\nu \int_\Omega |\nabla\omega|^2 d\mathbf{x}dz - 2\nu\gamma \int_\Gamma \int_z^\infty \omega\omega_z d\mathbf{x}dz \\
&\leq -\nu \int_\Omega |\nabla\omega|^2 d\mathbf{x}dz + \frac{\nu}{2}\int_\Omega |\omega_z|^2 d\mathbf{x}dz + 2\nu\gamma^2 \int_\Omega \omega^2 d\mathbf{x}dz \\
&\leq -\frac{\nu}{2}\int_\Omega |\nabla\omega|^2 d\mathbf{x}dz + 2\nu\gamma^2 \int_\Omega \omega^2 d\mathbf{x}dz,
\end{aligned} \tag{7.38}$$

where we have used the fact that the smooth solution of the heat equation ω decays to zero as $z \to \infty$. Thus, we get

$$\frac{1}{2}\frac{d}{dt}\int_\Omega \omega^2 d\mathbf{x}dz \leq 2\nu\gamma^2 \int_\Omega \omega^2 d\mathbf{x}dz.$$

It follows from Gronwall's inequality

$$e^{-4\nu\gamma^2 t}\int_\Omega \omega^2 d\mathbf{x}dz \leq \int_\Omega \omega_0^2 d\mathbf{x}dz = 0,$$

since $\omega_0 = 0$. Since $\omega \in H^s(\Omega)$ with $s > 3/2$, this implies that $\omega = 0$ for

$0 \le t < T$ which proves the uniqueness of smooth solutions for the heat equation with the mixed Dirichlet Robin boundary condition.

Next, we will prove the existence of the solution by constructing a solution explicitly. Let $\eta(\mathbf{x}, z, t)$ be the solution of the following initial boundary value problem:

$$\eta_t = \nu \Delta \eta, \quad (\mathbf{x}, z) \in \Omega,$$
$$\eta|_{\partial\Omega} = 0, \quad \eta|_{t=0} = \eta_0(\mathbf{x}, z),$$

and let $\xi(\mathbf{x}, t)$ be the solution of the following PDE in $\Omega_{\mathbf{x}}$:

$$\xi_t = \nu \Delta_{\mathbf{x}} \xi + \nu \gamma^2 \xi, \quad \mathbf{x} \in \Omega_{\mathbf{x}},$$
$$\xi|_{\partial\Omega_{\mathbf{x}}} = 0, \quad \xi|_{t=0} = \overline{\omega}_0(\mathbf{x}),$$

where $\Delta_{\mathbf{x}} = \frac{\partial^2}{\partial x_1^2} + \frac{\partial^2}{\partial x_2^2}$ and $\overline{\omega}_0(\mathbf{x}) = 2\gamma \int_0^\infty \omega_0(\mathbf{x}, z) e^{-\gamma z} dz$. From the standard theory of the heat equation, we know that η and ξ both exist globally in time.

We are interested in the case when the initial value $\eta_0(\mathbf{x}, z)$ is related to ω_0 by solving the following ODE as a function of z with $\mathbf{x}$ being fixed as a parameter:

$$-\frac{1}{\gamma}\eta_{0z} + \eta_0 = \omega_0(\mathbf{x}, z) - \overline{\omega}_0(\mathbf{x}) e^{-\gamma z}, \quad \eta_0(\mathbf{x}, 0) = 0. \tag{7.39}$$

Define

$$\omega(\mathbf{x}, z, t) \equiv -\frac{1}{\gamma}\eta_z + \eta + \xi(\mathbf{x}, t) e^{-\gamma z}, \quad (\mathbf{x}, z) \in \Omega. \tag{7.40}$$

It is easy to check that ω satisfies the heat equation for $t > 0$ and the initial condition. Obviously, ω also satisfies the boundary condition on $\partial\Omega \backslash \Gamma$. To verify the boundary condition on Γ, we observe by a direct calculation that $(\omega_z + \gamma\omega)|_\Gamma = -\frac{1}{\gamma}(\eta_z)_z|_\Gamma$. Since $\eta(\mathbf{x}, z)|_\Gamma = 0$, we obtain by using $\eta_t = \nu\Delta\eta$ and taking the limit as $z \to 0+$ that $\Delta\eta|_\Gamma = 0$, which implies that $\eta_{zz}|_\Gamma = 0$. Therefore, ω also satisfies the Dirichlet Robin boundary condition at Γ. This shows that ω is a solution of the system (7.33)–(7.35). By the uniqueness result that we proved earlier, the solution of the heat equation must be given by (7.40).

Since η and ξ are solutions of the heat equation with a standard Dirichlet boundary condition, the classical theory of the heat equation [15] gives the following regularity estimates:

$$\|\eta\|_{H^s(\Omega)} \le C \|\eta_0\|_{H^s(\Omega)}, \quad \|\xi(\mathbf{x})\|_{H^s(\Omega_{\mathbf{x}})} \le C e^{\nu\gamma^2 t} \|\overline{\omega}_0(\mathbf{x})\|_{H^s(\Omega_{\mathbf{x}})}.$$

Recall that $\eta_{zz}|_\Gamma = 0$. Therefore, η_z also solves the heat equation with the same Dirichlet Robin boundary condition:

$$(\eta_z)_t = \nu\Delta\eta_z, \quad (\mathbf{x}, z) \in \Omega,$$
$$(\eta_z)_z|_\Gamma = 0, \quad (\eta_z)|_{\partial\Omega\backslash\Gamma} = 0, \quad (\eta_z)|_{t=0} = \eta_{0z}(\mathbf{x}, z),$$

which implies that

$$\|\eta_z\|_{H^s(\Omega)} \leq C\|\eta_{0z}\|_{H^s(\Omega)}.$$

Putting all the above estimates for η, η_z and ξ together and using (7.40), we obtain the following estimate:

$$\begin{aligned}\|\omega\|_{H^s(\Omega)} &= \left\|-\frac{1}{\gamma}\eta_z + \eta + \xi(\mathbf{x}, t)e^{-\gamma z}\right\|_{H^s(\Omega)} \\ &\leq \frac{1}{\gamma}\|\eta_z\|_{H^s(\Omega)} + \|\eta\|_{H^s(\Omega)} + \left\|\xi(\mathbf{x}, t)e^{-\gamma z}\right\|_{H^s(\Omega)} \qquad (7.41) \\ &\leq C(\gamma, s)\left(\|\eta_{0z}\|_{H^s(\Omega)} + \|\eta_0\|_{H^s(\Omega)} + e^{\nu\gamma^2 t}\|\overline{\omega}_0(\mathbf{x})\|_{H^s(\Omega_\mathbf{x})}\right).\end{aligned}$$

It remains to bound $\|\eta_{0z}\|_{H^s(\Omega)}$, $\|\eta_0\|_{H^s(\Omega)}$ and $\|\overline{\omega}_0(\mathbf{x})\|_{H^s(\Omega_\mathbf{x})}$ in terms of $\|\omega_0\|_{H^s(\Omega)}$. By solving the ODE (7.39) directly, we can express η in terms of ω_0 explicitly

$$\eta_0(\mathbf{x}, z) = -\gamma e^{\gamma z}\int_0^z e^{-\gamma z'} f(\mathbf{x}, z')dz' = \gamma\int_z^\infty e^{-\gamma(z'-z)} f(\mathbf{x}, z')dz',$$

where $f(\mathbf{x}, z) = \omega_0(\mathbf{x}, z) - \overline{\omega}_0(\mathbf{x})e^{-\gamma z}$ and we have used the property that

$$\int_0^\infty f(\mathbf{x}, z)e^{-\gamma z}dz = 0.$$

By using integration by parts, we have

$$\begin{aligned}\eta_{0z}(\mathbf{x}, z) &= -\gamma f(\mathbf{x}, z) + \gamma^2\int_z^\infty e^{-\gamma(z'-z)} f(\mathbf{x}, z')dz' \\ &= \gamma\int_z^\infty e^{-\gamma(z'-z)} f_{z'}(\mathbf{x}, z')dz'.\end{aligned}$$

By induction we can show that for any $\alpha = (\alpha_1, \alpha_2, \alpha_3) \geq 0$

$$D^\alpha\eta_0 = \gamma\int_z^\infty e^{-\gamma(z'-z)} D^\alpha f(\mathbf{x}, z')dz'.$$

Let $K(z) = \gamma e^{-\gamma z}\chi(z)$ and $\chi(z)$ be the characteristic function

$$\chi(z) = \begin{cases} 0, & z \leq 0, \\ 1, & z > 0. \end{cases}$$

Then $D^\alpha \eta_0$ can be written in the following convolution form:

$$D^\alpha \eta_0(\mathbf{x}, z) = \int_0^\infty K(z' - z) D^\alpha f(\mathbf{x}, z') dz'.$$

Using Young's inequality (see e.g. page 232 of [17]), we obtain:

$$\begin{aligned} \|D^\alpha \eta_0\|_{L^2(\Omega)} &\leq \|K(z)\|_{L^1(\mathbb{R}^+)} \|D^\alpha f\|_{L^2(\Omega)} \\ &\leq C(\gamma) \left\| D^\alpha \omega_0 - (-\gamma)^{\alpha_3} e^{-\gamma z} D^{(\alpha_1, \alpha_2)} \overline{\omega}_0(\mathbf{x}) \right\|_{L^2(\Omega)} \qquad (7.42) \\ &\leq C(\gamma, \alpha) \left(\|D^\alpha \omega_0\|_{L^2(\Omega)} + \left\| D^{(\alpha_1, \alpha_2)} \overline{\omega}_0(\mathbf{x}) \right\|_{L^2(\Omega_{\mathbf{x}})} \right). \end{aligned}$$

Moreover, we obtain by using the Hölder inequality that

$$\begin{aligned} \left\| D^{(\alpha_1, \alpha_2)} \overline{\omega}_0(\mathbf{x}) \right\|_{L^2(\Omega_{\mathbf{x}})} &= \left(\int_{\Omega_{\mathbf{x}}} \left(\int_0^\infty e^{-\gamma z} D^{(\alpha_1, \alpha_2)} \omega_0(\mathbf{x}, z) dz \right)^2 d\mathbf{x} \right)^{1/2} \\ &\leq \left(\frac{1}{2\gamma} \int_{\Omega_{\mathbf{x}}} \int_0^\infty \left(D^{(\alpha_1, \alpha_2)} \omega_0(\mathbf{x}, z) \right)^2 dz d\mathbf{x} \right)^{1/2} \\ &= \frac{1}{\sqrt{2\gamma}} \left\| D^{(\alpha_1, \alpha_2)} \omega_0(\mathbf{x}, z) \right\|_{L^2(\Omega)}. \qquad (7.43) \end{aligned}$$

Substituting (7.43) into (7.42) yields

$$\|D^\alpha \eta_0\|_{L^2(\Omega)} \leq C(\gamma, \alpha) \left(\|D^\alpha \omega_0\|_{L^2(\Omega)} + \left\| D^{(\alpha_1, \alpha_2)} \omega_0 \right\|_{L^2(\Omega)} \right),$$

which implies that

$$\|\eta_0\|_{H^s(\Omega)} \leq C(\gamma, s) \|\omega_0\|_{H^s(\Omega)}, \quad \forall\, s \geq 0.$$

It follows from (7.43) that

$$\|\overline{\omega}_0(\mathbf{x})\|_{H^s(\Omega_{\mathbf{x}})} \leq C(\gamma) \|\omega_0\|_{H^s(\Omega)}, \quad \forall\, s \geq 0. \qquad (7.44)$$

On the other hand, we obtain from the equation for η_0 (7.39) that

$$\|\eta_{0z}\|_{H^s(\Omega)} = \gamma \|f + \eta_0\|_{H^s(\Omega)} \leq C(\gamma, s) \|\omega_0\|_{H^s(\Omega)}, \quad \forall\, s \geq 0. \quad (7.45)$$

Upon substituting (7.44)–(7.45) into (7.41), we obtain

$$\|\omega\|_{H^s(\Omega)} \leq C(\gamma, s) e^{\nu \gamma^2 t} \|\omega_0\|_{H^s(\Omega)}, \qquad (7.46)$$

where $C(\gamma, s)$ is a constant depending only on γ and s. This proves (7.36).

To prove (7.37), we use the classical regularity result for the heat equation with the homogeneous Dirichlet boundary condition to obtain the following estimates for $t > 0$:

$$\|\eta\|_{H^s(\Omega)} \leq C(t)\|\eta_0\|_{L^2(\Omega)}, \tag{7.47}$$

$$\|\eta_z\|_{H^s(\Omega)} \leq C(s,t)\|\eta_{0z}\|_{L^2(\Omega)}, \tag{7.48}$$

$$\|\overline{\omega}(\mathbf{x})\|_{H^s(\Omega_\mathbf{x})} \leq C(s,t)e^{\nu\gamma^2 t}\|\overline{\omega}_0(\mathbf{x})\|_{L^2(\Omega_\mathbf{x})}, \tag{7.49}$$

where $C(s,t)$ is a constant depending on s and t. By combining (7.47)–(7.49) with estimates (7.44)–(7.45), we obtain for any $t > 0$ that

$$\begin{aligned}\|\omega\|_{H^s(\Omega)} &\leq C(\gamma,s,t)\left(\|\eta_{0z}\|_{L^2(\Omega)} + \|\eta_0\|_{L^2(\Omega)} + e^{\nu\gamma^2 t}\|\overline{\omega}_0(\mathbf{x})\|_{L^2(\Omega_\mathbf{x})}\right)\\ &\leq C(\gamma,s,t)\|\omega_0\|_{L^2(\Omega)},\end{aligned}$$

where $C(\gamma,s,t) < \infty$ is a constant depending on γ, s and t. This proves (7.37) and completes the proof of the lemma. □

We also need the following well-known Sobolev inequality [18].

Lemma 7.10 *Let $u, v \in H^s(\Omega)$ with $s > 3/2$. We have*

$$\|uv\|_{H^s(\Omega)} \leq c\|u\|_{H^s(\Omega)}\|v\|_{H^s(\Omega)}.$$

Now we are ready to give the proof of Theorem 7.5.

Proof of Theorem 7.5 Let $v = u^2$. First, using the definition of the operator $\mathcal{K}$ (see Definition 7.7), we can rewrite the 3D model with partial viscosity in the following equivalent form:

$$\begin{cases} v_t &= 4v\mathcal{K}(\omega)_z \\ \omega_t &= v_z + \nu\Delta\omega \end{cases}, \quad (\mathbf{x},z) \in \Omega = \Omega_\mathbf{x} \times (0,\infty), \tag{7.50}$$

with the initial and boundary conditions given as follows:

$$\begin{aligned}&\omega|_{\partial\Omega\setminus\Gamma} = 0, \quad (\omega_z + \gamma\omega)|_\Gamma = 0,\\ &\omega|_{t=0} = \omega_0(\mathbf{x},z) \in W^s, \quad v|_{t=0} = v_0(\mathbf{x},z) \in V^{s+1},\end{aligned}$$

where $V^{s+1} = \{v \in H^{s+1} : v|_{\partial\Omega} = 0, v_z|_{\partial\Omega} = 0, v_{zz}|_{\partial\Omega} = 0\}$ and $W^s = \{w \in H^s : w|_{\partial\Omega\setminus\Gamma} = 0, (w_z + \gamma w)|_\Gamma = 0\}$.

We note that the condition $u_0|_{\partial\Omega} = u_{0z}|_{\partial\Omega} = 0$ implies that $v_0|_{\partial\Omega} = v_{0z}|_{\partial\Omega} = v_{0zz}|_{\partial\Omega} = 0$ by using the relation $v_0 = u_0^2$. Thus we have $v_0 \in V^{s+1}$. It is easy to show by using the u-equation that the property $u_0|_{\partial\Omega} = u_{0z}|_{\partial\Omega} = 0$ is preserved dynamically. Thus we have $v \in V^{s+1}$.

Define $U = (U_1, U_2) = (v, \omega)$ and $X = C([0,T]; V^{s+1}) \times C([0,T]; W^s)$ with the norm

$$\|U\|_X = \sup_{t\in[0,T]}\|U_1\|_{H^{s+1}(\Omega)} + \sup_{t\in[0,T]}\|U_2\|_{H^s(\Omega)}, \quad \forall U \in X$$

and let $S = \{U \in X : \|U\|_X \leq M\}$.

Now, define the map $\Phi : X \to X$ in the following way: let $\Phi(\tilde{v}, \tilde{\omega}) = (v, \omega)$. Then for any $t \in [0, T]$,

$$v(\mathbf{x}, z, t) = v_0(\mathbf{x}, z, t) + 4 \int_0^t \tilde{v}(\mathbf{x}, z, t')\mathcal{K}(\tilde{\omega})_z(\mathbf{x}, z, t')dt',$$
$$\omega(\mathbf{x}, z, t) = \mathcal{L}(\tilde{v}_z, \omega_0; \mathbf{x}, z, t),$$

where $\omega(\mathbf{x}, z, t) = \mathcal{L}(\tilde{v}_z, \omega_0; \mathbf{x}, z, t)$ is the solution of the following equation:

$$\omega_t = \tilde{v}_z + \nu \Delta \omega, \quad (\mathbf{x}, z) \in \Omega = \Omega_{\mathbf{x}} \times (0, \infty),$$

with the initial and boundary conditions:

$$\omega|_{\partial\Omega \backslash \Gamma} = 0, \quad (\omega_z + \gamma\omega)\,|_\Gamma = 0, \quad \omega|_{t=0} = \omega_0(\mathbf{x}, z).$$

We use the map Φ to define a Picard iteration: $U^{k+1} = \Phi(U^k)$ with $U^0 = (v_0, \omega_0)$. In the following, we will prove that there exist $T > 0$ and $M > 0$ such that

1. $U^k \in S$, for all k.
2. $\left\|U^{k+1} - U^k\right\|_X \leq \frac{1}{2}\left\|U^k - U^{k-1}\right\|_X$, for all k.

Then by the contraction mapping theorem, there exists $U = (v, \omega) \in S$ such that $\Phi(U) = U$ which implies that U is a local solution of the system (7.50) in X.

First, by Duhamel's principle, we have for any $g \in C([0, T]; V^s)$ that

$$\mathcal{L}(g, \omega_0; \mathbf{x}, z, t) = \mathcal{P}(\omega_0; 0, t) + \int_0^t \mathcal{P}(g; t', t)dt', \tag{7.51}$$

where $\mathcal{P}(g; t', t) = \tilde{g}(\mathbf{x}, z, t)$ is defined as the solution of the following initial boundary value problem at time t:

$$\tilde{g}_t = \nu \Delta \tilde{g}, \quad (\mathbf{x}, z) \in \Omega = \Omega_{\mathbf{x}} \times (0, \infty),$$

with the initial and boundary conditions:

$$\tilde{g}|_{\partial\Omega \backslash \Gamma} = 0, \quad (\tilde{g}_z + \gamma\tilde{g})\,|_\Gamma = 0, \quad \tilde{g}(\mathbf{x}, z, t') = g(\mathbf{x}, z, t').$$

We observe that $g(\mathbf{x}, z, t')$ also satisfies the same boundary condition as ω for any $0 \leq t' \leq t$ since $g = v_z^k$ and $v^k \in V^{s+1}$.

Now we can apply Lemma 7.8 to conclude that for any $t' < T$ and $t \in [t', T]$ we have

$$\|\mathcal{P}(g; t', t)\|_{H^s(\Omega)} \leq C(\gamma, s)e^{\nu\gamma^2(t-t')}\|g(\mathbf{x}, z, t')\|_{H^s(\Omega)},$$

which implies the following estimate for $\mathcal{L}$: for all $t \in [0,T]$,

$$\begin{aligned} &\|\mathcal{L}(g,\omega_0;\mathbf{x},z,t)\|_{H^s(\Omega)} \\ &\le C(\gamma,s)e^{\nu\gamma^2 t}\left(\|\omega_0\|_{H^s(\Omega)} + t\sup_{t'\in[0,t]}\|g(\mathbf{x},z,t')\|_{H^s(\Omega)}\right). \end{aligned} \tag{7.52}$$

Further, by using Lemma 7.6 and the above estimate (7.52) for the sequence $U^k = (v^k,\omega^k)$, we get the following estimate: $\forall t \in [0,T]$,

$$\begin{aligned} \left\|v^{k+1}\right\|_{H^{s+1}(\Omega)} &\le \|v_0\|_{H^{s+1}(\Omega)} + 4T\sup_{t\in[0,T]}\left\|v^k(\mathbf{x},z,t)\right\|_{H^{s+1}(\Omega)} \\ &\qquad\times\sup_{t\in[0,T]}\left\|\mathcal{K}(\omega^k)_z(\mathbf{x},z,t)\right\|_{H^{s+1}(\Omega)}, \\ &\le \|v_0\|_{H^{s+1}(\Omega)} + 4T\sup_{t\in[0,T]}\left\|v^k(\mathbf{x},z,t)\right\|_{H^{s+1}(\Omega)} \\ &\qquad\times\sup_{t\in[0,T]}\left\|\omega^k(\mathbf{x},z,t)\right\|_{H^s(\Omega)}, \end{aligned} \tag{7.53}$$

$$\begin{aligned} \left\|\omega^{k+1}\right\|_{H^s(\Omega)} &\le C(\gamma,s)e^{\nu\gamma^2 t}\left(\|\omega_0\|_{H^s(\Omega)} + t\sup_{t'\in[0,t]}\left\|v_z^k(\mathbf{x},z,t')\right\|_{H^s(\Omega)}\right) \\ &\le C(\gamma,s)e^{\nu\gamma^2 T}\left((\|\omega_0\|_{H^s(\Omega)} + T\sup_{t\in[0,T]}\left\|v^k\right\|_{H^{s+1}(\Omega)}\right). \end{aligned} \tag{7.54}$$

Next, we will use mathematical induction to prove that if T satisfies the following inequality:

$$8C(\gamma,s)Te^{\nu\gamma^2 T}\left(\|\omega_0\|_{H^s(\Omega)} + 2T\|v_0\|_{H^{s+1}(\Omega)}\right) \le 1 \tag{7.55}$$

then for all $k \ge 0$ and $t \in [0,T]$, we have that

$$\left\|v^k\right\|_{H^{s+1}(\Omega)} \le 2\|v_0\|_{H^{s+1}(\Omega)}, \tag{7.56}$$

$$\left\|\omega^k\right\|_{H^s(\Omega)} \le C(\gamma,s)e^{\nu\gamma^2 T}\left(\|\omega_0\|_{H^s(\Omega)} + 2T\|v_0\|_{H^{s+1}(\Omega)}\right). \tag{7.57}$$

First of all, $U^0 = (v_0,\omega_0)$ satisfies (7.56) and (7.57). Assume $U^k = (v^k,\omega^k)$ has this property, then for $U^{k+1} = (v^{k+1},\omega^{k+1})$, using (7.53) and (7.54), we have

$$\begin{aligned}\left\|v^{k+1}\right\|_{H^{s+1}(\Omega)} &\leq \|v_0\|_{H^{s+1}(\Omega)} + 4T \sup_{t\in[0,T]} \left\|v^k(\mathbf{x},z,t)\right\|_{H^{s+1}(\Omega)} \\ &\qquad \times \sup_{t\in[0,T]} \left\|\omega^k(\mathbf{x},z,t)\right\|_{H^s(\Omega)} \\ &\leq \|v_0\|_{H^{s+1}(\Omega)} \Big(1 + 8C(\gamma,s)Te^{\nu\gamma^2 T} \\ &\qquad \times \left(\|\omega_0\|_{H^s(\Omega)} + 2T\|v_0\|_{H^{s+1}(\Omega)}\right)\Big) \\ &\leq 2\|v_0\|_{H^{s+1}(\Omega)}, \quad \forall t\in[0,T],\end{aligned}$$

$$\begin{aligned}\left\|\omega^{k+1}\right\|_{H^s(\Omega)} &\leq C(\gamma,s)e^{\nu\gamma^2 T}\left(\|\omega_0\|_{H^s(\Omega)} + T\sup_{t\in[0,T]}\left\|v^k\right\|_{H^{s+1}(\Omega)}\right) \\ &\leq C(\gamma,s)e^{\nu\gamma^2 T}\left(\|\omega_0\|_{H^s(\Omega)} + 2T\|v_0\|_{H^{s+1}(\Omega)}\right),\end{aligned}$$

$\forall t \in [0,T]$. Then, by induction, we prove that for any $k \geq 0$, $U^k = (v^k, \omega^k)$ is bounded by (7.56) and (7.57).

We want to point out that there exists $T > 0$ such that the inequality (7.55) is satisfied. One choice of T is given as following:

$$T_1 = \min\left\{\left[8C(\gamma,s)e^{\nu\gamma^2}\left(\|\omega_0\|_{H^s(\Omega)} + 2\|v_0\|_{H^{s+1}(\Omega)}\right)\right]^{-1}, 1\right\}. \quad (7.58)$$

Using the choice of T in (7.58), we can choose

$$M = 2\|v_0\|_{H^{s+1}(\Omega)} + C(\gamma,s)e^{\nu\gamma^2}\left(\|\omega_0\|_{H^s(\Omega)} + 2\|v_0\|_{H^{s+1}(\Omega)}\right).$$

Then we have $U^k \in S$, for all k.

Next, we will prove that Φ is a contraction mapping for some small $0 < T \leq T_1$. First of all, by definition of the iteration scheme we have

$$\begin{aligned}&\left\|v^{k+1} - v^k\right\|_{H^{s+1}(\Omega)} \\ &= \Big\|\int_0^t v^k(\mathbf{x},t')\mathcal{K}(\omega^k)_z(\mathbf{x},t')dt' \\ &\qquad - \int_0^t v^{k-1}(\mathbf{x},t')\mathcal{K}(\omega^{k-1})_z(\mathbf{x},t')dt'\Big\|_{H^{s+1}(\Omega)}.\end{aligned}$$

By the triangle inequality this is dominated by

$$\left\| \int_0^t \left(v^k - v^{k-1}\right)(\mathbf{x},t')\mathcal{K}(\omega^k)_z(\mathbf{x},t')dt' \right\|_{H^{s+1}(\Omega)} + \left\| \int_0^t v^{k-1}(\mathbf{x},t')\left(\mathcal{K}(\omega^k)_z - \mathcal{K}(\omega^{k-1})_z\right)(\mathbf{x},t')dt' \right\|_{H^{s+1}(\Omega)}.$$

By Lemmas 7.6 and 7.10, the above is bounded by

$$\begin{aligned} &T \sup_{t\in[0,T]} \left\|v^k - v^{k-1}\right\|_{H^{s+1}(\Omega)} \sup_{t\in[0,T]} \left\|\mathcal{K}(\omega^k)_z\right\|_{H^{s+1}(\Omega)} \\ &\qquad + T \sup_{t\in[0,T]} \left\|v^{k-1}\right\|_{H^{s+1}(\Omega)} \sup_{t\in[0,T]} \left\|\mathcal{K}(\omega^k - \omega^{k-1})_z\right\|_{H^{s+1}(\Omega)} \\ &\le MT\left(\sup_{t\in[0,T]} \left\|v^k - v^{k-1}\right\|_{H^{s+1}(\Omega)} + \sup_{t\in[0,T]} \left\|\omega^k - \omega^{k-1}\right\|_{H^s(\Omega)}\right). \end{aligned}$$

On the other hand, Lemma 7.8 and (7.51) imply

$$\begin{aligned} \left\|\omega^{k+1} - \omega^k\right\|_{H^s(\Omega)} &= \left\|\mathcal{L}(v_z^k, \omega_0; \mathbf{x}, t) - \mathcal{L}(v_z^{k-1}, \omega_0; \mathbf{x}, t)\right\|_{H^s(\Omega)} \\ &\le \left\| \int_0^t \mathcal{P}(v_z^k - v_z^{k-1}; t', t)dt' \right\|_{H^s(\Omega)} \\ &\le TC(\gamma, s)e^{\nu\gamma^2 T} \sup_{t\in[0,T]} \left\|v_z^k - v_z^{k-1}\right\|_{H^s(\Omega)} \\ &\le TC(\gamma, s)e^{\nu\gamma^2 T} \sup_{t\in[0,T]} \left\|v^k - v^{k-1}\right\|_{H^{s+1}(\Omega)}. \end{aligned}$$

Let

$$\begin{aligned} T = \min\Bigg\{ &\left[8C(\gamma, s)e^{\nu\gamma^2}\left(\|\omega_0\|_{H^s(\Omega)} + 2\|v_0\|_{H^{s+1}(\Omega)}\right)\right]^{-1}, \\ &\left[2C(\gamma, s)e^{\nu\gamma^2}\right]^{-1}, \frac{1}{2M}, 1\Bigg\}. \end{aligned} \tag{7.59}$$

Then, we have

$$\left\|U^{k+1} - U^k\right\|_X \le \frac{1}{2}\left\|U^k - U^{k-1}\right\|_X.$$

This proves that the sequence U^k converges to a fixed point of the map $\Phi : X \to X$, and the limiting fixed point $U = (v, \omega)$ is a solution of the 3D model with partial viscosity. Moreover, by passing to the limit in (7.56)–(7.57), we obtain the following *a priori* estimate for the solution

(v, ω):

$$\|v\|_{H^{s+1}(\Omega)} \le 2\|v_0\|_{H^{s+1}(\Omega)},$$
$$\|\omega\|_{H^s(\Omega)} \le C(\gamma, s)e^{\nu\gamma^2 T}\left(\|\omega_0\|_{H^s(\Omega)} + 2T\|v_0\|_{H^{s+1}(\Omega)}\right),$$

for $0 \le t \le T$ with T defined in (7.59).

It remains to show that the smooth solution of the 3D model with partial viscosity is unique. Let (v_1, ω_1) and (v_2, ω_2) be two smooth solutions of the 3D model with the same initial data and satisfying $\|v_i\|_{H^{s+1}(\Omega)} \le M$ and $\|\omega_i\|_{H^s(\Omega)} \le M$ for $i = 1, 2$ and $0 \le t \le T$, where M is a positive constant depending on the initial data as well as γ, s, and T. Since $s > 3/2$, the Sobolev embedding theorem [15] implies that

$$\|v_i\|_{L^\infty(\Omega)} \le \|v_i\|_{H^{s+1}(\Omega)} \le M, \quad i = 1, 2, \tag{7.60}$$

$$\begin{aligned}\|\mathcal{K}(\omega_i)_z\|_{L^\infty(\Omega)} &\le \|\mathcal{K}(\omega_i)_z\|_{H^s(\Omega)} \\ &\le C_s\|\omega_i\|_{H^s(\Omega)} \le C_s M, \quad i = 1, 2.\end{aligned} \tag{7.61}$$

Let $v = v_1 - v_2$ and $\omega = \omega_1 - \omega_2$. Then (v, ω) satisfies

$$\left\{\begin{array}{lcl} v_t &=& 4v\mathcal{K}(\omega_1)_z + 4v_2\mathcal{K}(\omega)_z \\ \omega_t &=& v_z + \nu\Delta\omega \end{array}\right., \quad (\mathbf{x}, z) \in \Omega = \Omega_{\mathbf{x}} \times (0, \infty),$$

with $\omega|_{\partial\Omega\setminus\Gamma} = 0$, $(\omega_z + \gamma\omega)|_\Gamma = 0$, and $\omega|_{t=0} = 0$, $v|_{t=0} = 0$. By using (7.60)–(7.61), and proceeding as the uniqueness estimate for the heat equation in (7.38), we can derive the following estimate for v and ω:

$$\frac{d}{dt}\|v\|^2_{L^2(\Omega)} \le C_1(\|v\|^2_{L^2(\Omega)} + \|\omega\|^2_{L^2(\Omega)}), \tag{7.62}$$

$$\frac{d}{dt}\|\omega\|^2_{L^2(\Omega)} \le C_3(\|v\|^2_{L^2(\Omega)} + \|\omega\|^2_{L^2(\Omega)}), \tag{7.63}$$

where C_i $(i = 1, 2, 3)$ are positive constants depending on M, ν, γ, C_s. In obtaining the estimate for (7.63), we have performed integration by parts in the estimate of the v_z-term in the ω-equation and absorbing the contribution from ω_z by the diffusion term. There is no contribution from the boundary term since $v|_{z=0} = 0$. We have also used the property $\|\mathcal{K}(\omega)_z\|_{L^2(\Omega)} \le C_s\|\omega\|_{L^2(\Omega)}$, which can be proved directly by following the argument in the Appendix of [26]. Since $v_0 = 0$ and $\omega_0 = 0$, the Gronwall inequality implies that $\|v\|_{L^2(\Omega)} = \|\omega\|_{L^2(\Omega)} = 0$ for $0 \le t \le T$. Furthermore, since $v \in H^{s+1}$ and $\omega \in H^s$ with $s > 3/2$, v and ω are continuous. Thus we must have $v = \omega = 0$ for $0 \le t \le T$. This proves the uniqueness of the smooth solution for the 3D model. □

Acknowledgments. Hou's work was in part supported by the NSF through the grant DMS-0908546 and an AFOSR MURI grant FA9550-09-1-0613.

References

[1] J. T. Beale, T. Kato and A. Majda, Remarks on the breakdown of smooth solutions for the 3-D Euler equations. *Comm. Math. Phys.* **94**, 61–66, 1984.

[2] L. Caffarelli, R. Kohn and L. Nirenberg, Partial regularity of suitable weak solutions of the Navier-Stokes equations. *Comm. Pure Appl. Math.*, **35**, 771–831, 1982.

[3] R. Caflisch and M. Siegel, A semi-analytic approach to Euler singularities. *Methods and Appl. of Analysis*, **11**, 423–430, 2004.

[4] D. Chae, A. Cordoba, D., Cordoba and M. A. Fontelos, Finite time singularities in a 1D model of the quasi-geostrophic equation. *Adv. Math.*, **194**, 203–223, 2005.

[5] P. Constantin, Note on loss of regularity for solutions of the 3D incompressible Euler and related equations. *Commun. Math. Phys.*, **104**, 311–326, 1986.

[6] P. Constantin, C. Fefferman and A. Majda, Geometric constraints on potentially singular solutions for the 3-D Euler equation. *Commun. in PDEs*, **21**, 559–571, 1996.

[7] D. Cordoba and C. Fefferman, Growth of solutions for QG and 2D Euler equations. *J. Amer. Math. Soc.*, **15**, 665–670 (electronic), 2002.

[8] A. Cordoba, D. Cordoba and M. A. Fontelos, Formation of singularities for a transport equation with nonlocal velocity. *Ann. of Math.*, **162**, 1–13, 2005.

[9] P. Constantin, P. D. Lax and A. J. Majda, A simple one-dimensional model for the three-dimensional vorticity equation. *Comm. Pure Appl. Math.*, **38**, 715–724, 1985.

[10] J. Deng, T. Y. Hou and X. Yu, Geometric properties and non-blowup of 3-D incompressible Euler flow. *Commun. PDEs*, **30**, 225–243, 2005.

[11] J. Deng, T. Y. Hou and X. Yu, Improved geometric conditions for non-blowup of 3D incompressible Euler equation. *Commun. PDEs*, **31**, 293–306, 2006.

[12] S. De Gregorio, On a one-dimensional model for the 3-dimensional vorticity equation. *J. Stat. Phys.*, **59**, 1251–1263, 1990.

[13] S. De Gregorio, A partial differential equation arising in a 1D model for the 3D vorticity equation. *Math. Method Appl. Sci.*, **19**, 1233–1255, 1996.

[14] R. J. DiPerna and P. L. Lions On the Cauchy problem for Boltzmann equations: global existence and weak stability. *Ann. Math.*, **130**, 321–366, 1989.

[15] L. C. Evans, *Partial Differential Equations.* American Mathematical Society Publ., 1998.

[16] C. Fefferman, *http://www.claymath.org/millennium/Navier-Stokes equations.*

[17] G. B. Foland, *Real Analysis – Modern Techniques and Their Applications.* Wiley-Interscience Publ., Wiley and Sons, Inc., 1984.

[18] G. B. Foland, *Introduction to Partial Differential Equations.* Princeton University Press, Princeton, N.J., 1995.

[19] T. Y. Hou and R. Li, Dynamic depletion of vortex stretching and non-blowup of the 3-D incompressible Euler equations. *J. Nonlinear Science,* **16**, 639–664, 2006.

[20] T. Y. Hou and R. Li, Computing nearly singular solutions using pseudo-spectral methods. *J. Comput. Phys.*, **226**, 379–397, 2007.

[21] T. Y. Hou and R. Li, Blowup or no blowup? The interplay between theory and numerics. *Physica D*, **237**, 1937–1944, 2008.

[22] T. Y. Hou and C. Li, Dynamic stability of the 3D axi-symmetric Navier-Stokes equations with swirl. *Comm. Pure Appl. Math.*, **61**, 661–697, 2008.

[23] T. Y. Hou and Z. Lei, On the stabilizing effect of convection in 3D incompressible flows. *Comm. Pure Appl. Math.*, **62**, 501–564, 2009.

[24] T. Y. Hou and Z. Lei, On partial regularity of a 3D model of Navier-Stokes equations. *Commun. Math Phys.*, **287**, 281–298, 2009.

[25] T. Y. Hou, C. Li, Z. Shi, S. Wang, and X. Yu, On singularity formation of a nonlinear nonlocal system. *Arch. Ration. Mech. Anal.*, **199**, 117–144, 2011.

[26] T. Y. Hou, Z. Shi and S. Wang On singularity formation of a 3D model for incompressible Navier-Stokes equations. arXiv:0912.1316v1 [math.AP], submitted to *Adv. Math.*.

[27] T. Y. Hou and Z. Shi, Dynamic growth estimates of maximum vorticity for 3D incompressible Euler equations and the SQG model. *DCDS-A*, **32**, 2011.

[28] R. Kerr, Evidence for a singularity of the three dimensional, incompressible Euler equations. *Phys. Fluids*, **5**, 1725–1746, 1993.

[29] R. Kerr, Velocity and scaling of collapsing Euler vortices. *Phys. Fluids,* **17**, 075103-114, 2005.

[30] D. Li and J. Rodrigo, Blow up for the generalized surface quasi-geostrophic equation with supercritical dissipation. *Comm. Math. Phys.*, **286**, 111–124, 2009.

[31] D. Li and Y. G. Sinai, Blow ups of complex solutions of the 3D Navier-Stokes system and renormalization group method. *J. Europ. Math. Soc.*, **10**, 267–313, 2008.

[32] A. J. Majda and A. L. Bertozzi, *Vorticity and Incompressible Flow.* Cambridge Texts in Applied Mathematics, 27. Cambridge University Press, Cambridge, 2002.

[33] T. Matsumotoa, J. Bec and U. Frisch, Complex-space singularities of 2D Euler flow in Lagrangian coordinates. *Physica D*, **237**, 1951–1955, 2007.

[34] G. Prodi, Un teorema di unicità per le equazioni di Navier-Stokes. *Ann. Mat. Pura Appl.*, **48**, 173–182, 1959.

[35] J. Serrin, The initial value problem for the Navier-Stokes equations. In *Nonlinear Problems*, Univ. of Wisconsin Press, Madison, 69–98, 1963.

[36] R. Temam, *Navier-Stokes Equations.* Second Edition, AMS Chelsea Publishing, Providence, RI, 2001.

8

Algebraic and Differential Invariants

Evelyne Hubert
INRIA Méditérranée
Sophia Antipolis

Abstract

This article highlights a coherent series of algorithmic tools to compute and work with algebraic and differential invariants.

8.1 Introduction

Group actions are ubiquitous in mathematics and arise in diverse fields of science and engineering, including physics, mechanics, and computer vision. Invariants of these group actions typically arise to reduce a problem or to decide if two objects, geometric or abstract, are obtained from one another by the action of a group element. [8, 9, 10, 11, 13, 15, 17, 39, 40, 42, 43, 45, 46, 52, 59] are a few recent references of applications. Both algebraic and differential invariant theories have become in recent years the subject of computational mathematics [13, 14, 17, 40, 60]. Algebraic invariant theory studies polynomial or rational invariants of algebraic group actions [18, 22, 23, 54]. A typical example is the discriminant of a quadratic binary form as an invariant of an action of the special linear group. The differential invariants appearing in differential geometry are smooth functions on a jet bundle that are invariant under a prolonged action of a Lie group [4, 16, 34, 48, 53]. A typical example is the curvature of a plane curve, invariant under the action of the group of the isometries on the plane. Curvature is not a rational function, but an algebraic function. Concomitantly the classical Lie groups are linear algebraic groups.

This article reviews results of [14, 28, 29, 30, 31, 32] in order to show their coherence in addressing algorithmically an algebraic description of the differential invariants of a group action. In the first section we show how to compute the rational invariants of a group action and give concrete expressions to a set of algebraic invariants that are of fundamental importance in the differential context. The second section addresses the question of finite representation of differential invariants with invariant derivations, a set of generating differential invariants and the differential relationships among them. In the last section we describe the algebraic structure that better serves the representation of differential invariants.

8.2 Algebraic invariants

This section offers a geometric description and the algebraic computation of the invariants of a group action. In the first subsection we provide the geometric description of *normalized invariants* of a Lie group action. This description, based solely on the concept of cross-section, is directly drawn from [32] and is an alternative to the moving frame based description of [14, 40]. In the second subsection we provide an algorithm to compute a generating set of rational invariants for the rational action of an algebraic group. This is a quick presentation of the main results of [31]. In the last subsection we explain how the algorithm also delivers the normalized invariants as algebraic invariants. The details are to be found in [32].

The algorithm to compute rational invariants relies on Gröbner bases. This is a founding stone in algebraic computing. Introduced in the mid 60s they have been the subject of intensive research and now textbooks [1, 3, 12, 20, 37]. They are implemented in most computer algebra systems. Among their many applications, they algorithmically compute the elimination ideals that appear in the algorithmic construction of a generating set of rational invariants.

8.2.1 A geometric vision

We consider a Lie group $\mathcal{G}$, with identity denoted by e and dimension r, and a smooth manifold $\mathcal{Z}$ of dimension n. An action of $\mathcal{G}$ on $\mathcal{Z}$ is given by a smooth map $\mathcal{G} \times \mathcal{Z} \to \mathcal{Z}$. The image of a pair (λ, z) is denoted $\lambda \star z$. If $\lambda \cdot \mu$ denotes the product of two elements $\lambda, \mu \in \mathcal{G}$, the action satisfies $\mu \star (\lambda \star z) = (\mu \cdot \lambda) \star z$ and $e \star z = z$ for all $z \in \mathcal{Z}$. To be of practical interest, the notion of an action is often relaxed to being defined on an open subset of $\mathcal{G} \times \mathcal{Z}$ that contains $\{e\} \times \mathcal{Z}$. All the subsequent relationships are understood by restricting them to the locus where the quantities appearing are well defined.

The *orbit* $\mathcal{O}_z$ of a point $z \in \mathcal{Z}$ is the set of points that are the image of z by some $\lambda \in \mathcal{G}$: $\mathcal{O}_z = \{\lambda \star z \,|\, \lambda \in \mathcal{G}\}$. A (global) *cross-section* is an embedded submanifold $\mathcal{P}$ of $\mathcal{Z}$ that intersects each orbit of $\mathcal{Z}$ at a unique point. The *invariants* of the action are the functions that are constant on orbits: They satisfy $f(\lambda \star z) = f(z)$. If a cross-section $\mathcal{P}$ exists, one easily understands that functions on this cross-section are in one-to-one correspondence with invariants. An invariant defines a function on $\mathcal{P}$ by restriction. Conversely, each function $\bar{f} : \mathcal{P} \to \mathbb{R}$ on the cross-section defines an invariant $f : \mathcal{Z} \to \mathbb{R}$ by spreading its value along the orbit: $f(z) = \bar{f}(\bar{z})$ where $\bar{z}$ is the intersection of $\mathcal{O}_z$ with $\mathcal{P}$. Going further with this idea we are led to define the *invariantization* $\bar{\iota} f$ of a function f on $\mathcal{Z}$: $\bar{\iota} f$ is the unique invariant that agrees with f on the cross-section: $\bar{\iota} f(z) = f(\bar{z})$ where $\bar{z}$ is the intersection of $\mathcal{O}_z$ with $\mathcal{P}$. We thus retain the values of f on $\mathcal{P}$ and spread them along the orbit.

The global picture we just drew can not be easily put into practice as the existence of a global cross-section is not secured, not to mention the difficulty of identifying one if it does exist. Yet if we accept the idea of restricting to the neighborhood of a point where the action is well-behaved, things unravel pretty nicely. Furthermore infinitesimal calculus comes into the picture to help. And indeed the faithful description of Lie group action by its infinitesimal generators has made it successful in applications [47].

The tangent space $T\mathcal{G}|_e$ can be identified with the Lie algebra $\mathfrak{g}$ of $\mathcal{G}$. To every vector $\hat{v}$ in $T\mathcal{G}|_e$ we can associate a smooth vector field v on $\mathcal{Z}$. The integral curves of this vector field are the orbits of a one-dimensional subgroup of $\mathcal{G}$. Such a vector field is called an *infinitesimal generator* and is often understood as a derivation: For a function f on $\mathcal{Z}$, $\mathrm{v}(f)$ measures the variation of f along the integral curve. If f is an invariant of the group action then $\mathrm{v}(f) = 0$.

If $\hat{v}_1, \ldots, \hat{v}_r$ is a basis for the Lie algebra of $\mathcal{G}$, then the associated infinitesimal generators $\mathrm{v}_1, \ldots, \mathrm{v}_r$ span the tangent space to the orbits at each point of $\mathcal{Z}$. The dimension of the orbit of a point z is the rank of $\mathrm{v}_1, \ldots, \mathrm{v}_r$ at z. We shall place ourselves in a neighborhood $\mathcal{U}$ of a point z where this rank is constant and equal to d. Obviously $d \leq r$, the dimension of the group. *Local invariants* are those functions f on $\mathcal{Z}$ for which $\mathrm{v}_1(f) = 0, \ldots, \mathrm{v}_r(f) = 0$ in this neighborhood.

By possibly further restricting the neighborhood $\mathcal{U}$, we can then prove the existence of a *local cross-section* $\mathcal{P}$, that is an embedded submanifold of dimension $n - d$ that intersects transversally the connected part of the orbits of $\mathcal{U}$ at a single point. This submanifold can be described as the zero set of d independent functions $(p_1, \ldots, p_d)$ on $\mathcal{U}$. The condition that those functions define a cross-section is that the rank of the $r \times d$ matrix $V(P) = (\mathrm{v}_i(p_j))_{i=1..r}^{j=1..d}$ is d on $\mathcal{P}$.

One thus sees that a lot of freedom comes into the choice of a local cross-section. In particular, if $(z_1, \ldots, z_n)$ are coordinate functions on $\mathcal{U}$ one shows that we can choose a cross-section as the level set of d of these coordinate functions as a practical choice. The invariantization process we described earlier can be applied by restriction to $\mathcal{U}$ and the invariantization of the coordinate functions $(\bar{\iota}z_1, \ldots, \bar{\iota}z_n)$ are singled out as the *normalized invariants*. They functionally generate all local invariants and the equations of the cross-section describe completely their relationships. We formalize this here as a theorem, but we refer to [32] for a precise statement.

Theorem 8.1 *Let $\mathcal{P}$ be a local cross-section on $\mathcal{U}$, given as the zero set of d independent functions $p_1, \ldots, p_d$. The normalized invariants $(\bar{\iota}z_1, \ldots, \bar{\iota}z_n)$ of the coordinate functions $(z_1, \ldots, z_n)$ satisfy:*

- $p_1(\bar{\iota}z_1, \ldots, \bar{\iota}z_n) = 0, \ldots, p_d(\bar{\iota}z_1, \ldots, \bar{\iota}z_n) = 0,$
- *if a function p is such that $p(\bar{\iota}z_1, \ldots, \bar{\iota}z_n) = 0$ then there exist, around each point of the cross-section, functions $(a_1, \ldots, a_r)$ such that $p = a_1 p_1 + \cdots + a_d p_d$,*
- *if f is a local invariant then $f(z_1, \ldots, z_n) = f(\bar{\iota}z_1, \ldots, \bar{\iota}z_n)$.*

Example 8.2 We consider the linear action of $SO(2)$, the group of 2×2 orthogonal matrices with determinant 1, on $\mathbb{R}^2$. The action of an element of the group is a rotation with the origin as center. The orbits are the circles centered at the origin, and the origin itself.

The positive z_1-axis, $\mathcal{P} = \{(z_1, z_2) | z_2 = 0, z_1 > 0\}$, is a local cross-section on $\mathcal{Z}$. The invariantization of the coordinate functions are the functions $\bar{\iota}z_1$ and $\bar{\iota}z_2$ that associate to a point (z_1, z_2) the coordinates of the intersection of its orbit with the cross-section. Thus

$$\bar{\iota}z_1 : (z_1, z_2) \mapsto \sqrt{z_1^2 + z_2^2} \quad \text{and} \quad \bar{\iota}z_2 : (z_1, z_2) \mapsto 0.$$

By Theorem 8.1, all local invariants can be written in terms of $\sqrt{z_1^2 + z_2^2}$ by carrying out the substitutions $z_1 \to \sqrt{z_1^2 + z_2^2}$ and $z_2 \to 0$.

The above example is specific in as much as the dimension of the orbits, outside of the origin, is equal to the dimension of the group.

This case is of great importance in the differential context for which the presented geometric construction was first drawn in [14]. We can indeed then introduce the seminal notion of a moving frame. A moving frame $\rho : \mathcal{Z} \rightarrow \mathcal{G}$ is a smooth map to the group that is (right) equivariant, i.e. $\rho(\lambda \star z) = \rho(z) \cdot \lambda^{-1}$. When the dimension of the orbits is equal to the dimension of the group, a local cross-section determines a moving frame. For any point z in the neighbourhood of the cross-section we can single out an element λ of the group, close enough to the identity, such that $\lambda \star z$ belongs to the cross-section. The map that associates such a λ to a point z is a moving frame. If the cross-section is given as the zero set of the functions $p_1, \ldots, p_r$ then

$$p_1(\lambda \star z) = 0, \ldots, p_r(\lambda \star z) = 0$$

implicitly define this moving frame. Indeed, the transversality condition of the cross-section,

$$\det V(P) = \det \left(\mathrm{v}_i(p_j)\right)_{i=1..r}^{j=1..r} \neq 0$$

then ensures that we can apply the implicit function theorem. This actually provides another way of characterizing invariantization as indeed we then have $\bar{\iota} f(z) = f(\rho(z) \star z)$.

8.2.2 Algorithms for rational invariants

In this section $\mathbb{K}$ is a field of characteristic zero, typically $\mathbb{Q}$, $\mathbb{R}$ or even $\mathbb{C}$, while $\bar{\mathbb{K}}$ is an algebraically closed field that contains it. We consider an r-dimensional (affine) algebraic group given as the variety, in $\mathbb{K}^l$ or $\bar{\mathbb{K}}^l$, of an ideal G in $\mathbb{K}[\lambda] = \mathbb{K}[\lambda_1, \ldots, \lambda_l]$. The group operation and its inverse are defined by polynomial maps. A rational action of $\mathcal{G}$ on the affine space $\mathcal{Z} = \mathbb{K}^n$ can be given as an n-tuple of rational functions in $\mathbb{K}(\lambda_1, \ldots, \lambda_l, z_1, \ldots, z_n)$ provided they are well defined on $\{e\} \times \mathcal{Z} \subset \mathcal{G} \times \mathcal{Z}$.

There exists an invariant open subset $\mathcal{Z}_0 \subset \mathcal{Z}$ such that the orbits of the induced action of $\mathcal{G}$ on $\mathcal{Z}_0$ all have the same dimension, say d. This is the dimension of generic orbits. Rational invariants are the rational functions on $\mathcal{Z}$ that are constant on orbits of the action. They form a subfield $\mathbb{K}(z)^G$ of $\mathbb{K}(z)$, which is therefore finitely generated. The transcendence degree of $\mathbb{K}(z)^G$ over $\mathbb{K}$ is then $n - d$. A generating set of rational invariants thus has at least $n - d$ elements. If there are more than that, then some are algebraically dependent on the others.

These statements, that can be found in [54], can actually be recovered in a constructive way. Assume that $\lambda \star z = \left(\frac{g_1(\lambda,z)}{h_1(\lambda,z)}, \ldots, \frac{g_n(\lambda,z)}{h_n(\lambda,z)}\right)$ where $g_i, h_i \in \mathbb{K}[\lambda, z]$. Let h be the least common multiple of the denominators $h_1, \ldots, h_n$ and introduce a second set of variables $Z_1, \ldots, Z_n$. Consider the elimination ideal $O_z = (G + (h_1 Z_1 - g_1, \ldots, h_n Z_n - g_n)) : h^\infty \cap \mathbb{K}(z_1, \ldots, z_n)[Z_1, \ldots, Z_n]$, that we suggestively write as

$$O_z = (G + (Z - \lambda \star z)) \cap \mathbb{K}(z)[Z],$$

for short. The variety of O_z, for a generic $z \in \mathbb{K}^n$, is the Zariski closure of $\mathcal{O}_z$. For $\lambda \in \mathcal{G}$ and $z \in \mathcal{Z}$ we have $\mathcal{O}_z = \mathcal{O}_{\lambda \star z}$. The ideal O_z of $\mathbb{K}(z)[Z]$ is thus left unchanged when we substitute z by $\lambda \star z$. Any canonical rational representation of this ideal must thus have its coefficients in $\mathbb{K}(z)^G$. The coefficients of the Chow form of O_z were first shown to form a separating set of rational invariants, and hence a generating set [58]. In [31, 44, 35] a reduced Gröbner basis for O_z is used. Its coefficients are shown to form a generating set of rational invariants by exhibiting an algorithm to rewrite any other rational invariants in terms of those.

The distinguishing feature of [31] is to additionally incorporate the idea of the geometric construction reviewed in the previous section. In the present context, a *cross-section of degree e* is an irreducible subvariety $\mathcal{P}$ of $\mathcal{Z}$ that intersects the generic orbits in e distinct points. If $P \subset \mathbb{K}[Z]$ is a prime ideal of codimension d, its variety $\mathcal{P}$ is a cross-section if and only if the ideal

$$I_z = (G + (Z - \lambda \star z) + P) \cap \mathbb{K}(z)[Z]$$

is radical and zero-dimensional. This is equivalent to saying that the quotient $\mathbb{K}(z)[Z]/I_z$ is finite dimensional as a $\mathbb{K}(z)$-vector space. Its dimension e is then the degree of the cross-section $\mathcal{P}$ and is easily read on a Gröbner basis of I_z. As before $I_{\lambda \star z} = I_z$ and we obtain the same results as for O_z about the coefficients of a canonical representation. Let us give a precise statement [31, Theorem 2.16 and 3.7].

Theorem 8.3 *Consider $\{r_1, \ldots, r_\kappa\} \in \mathbb{K}(z)$ the coefficients of a reduced Gröbner basis Q of O_z or I_z. Then $\{r_1, \ldots, r_\kappa\}$ is a generating set of rational invariants: $\mathbb{K}(z)^G = \mathbb{K}(r_1, \ldots, r_\kappa)$. Furthermore we can rewrite any rational invariant $\frac{p}{q}$, with $p, q \in \mathbb{K}[z]$, in terms of those as follows.*

Take a new set of indeterminates $y_1, \ldots, y_\kappa$ and consider the set $Q_y \subset \mathbb{K}[y, Z]$ obtained from Q by substituting each of the r_i by y_i. Let $a(y, Z) = \sum_{\alpha \in \mathbb{N}^n} a_\alpha(y) Z^\alpha$ and $b(y, Z) = \sum_{\alpha \in \mathbb{N}^n} b_\alpha(y) Z^\alpha$ in $\mathbb{K}[y, Z]$ be

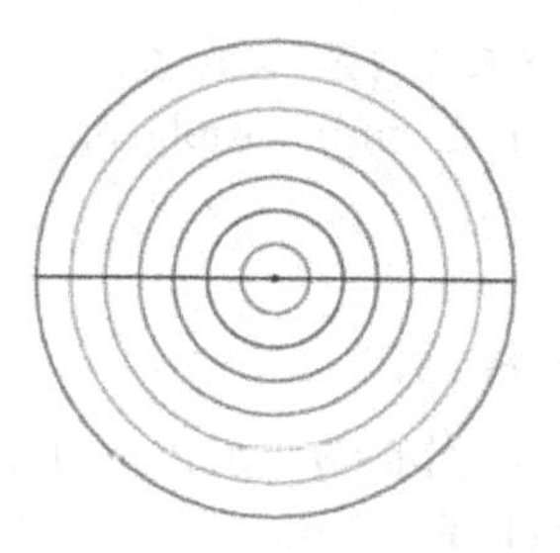

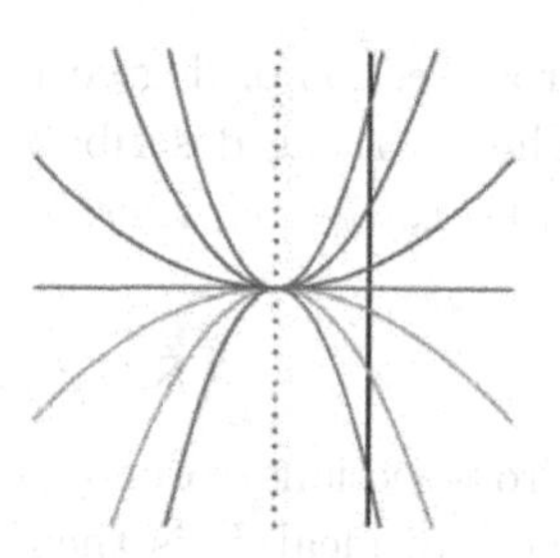

Figure 8.1 Left: The orbits and a cross-section of the group of rotations (Example 8.4); Right: The orbits of the action of Example 8.5.

the reductions of $p(Z)$ and $q(Z)$ with respect to Q_y. There exists $\alpha \in \mathbb{N}^n$ such that $b_\alpha(r) \neq 0$ and for any such α we have $\frac{p(z)}{q(z)} = \frac{a_\alpha(r)}{b_\alpha(r)}$.

The first advantage of using I_z instead of O_z is that zero dimensional ideals are more amenable when it comes to computing Gröbner bases. The second advantage that cannot be underestimated, though it does not come forth in the too simple examples we examine below, is that the output is considerably smaller: There are many less non-trivial coefficients in the Gröbner basis of I_z than in that of O_z. We thus have a smaller set of generators.

Example 8.4 We consider the group of rotation of the plane. The group is defined as the circle and given by the ideal $G = (\lambda_1^2 + \lambda_2^2 - 1) \subset \mathbb{K}[\lambda_1, \lambda_2]$. Its linear action on $\mathbb{K}^2$ is given by:

$$\lambda \star z = (\lambda_1 z_1 - \lambda_2 z_2,\ \lambda_2 z_1 + \lambda_1 z_2).$$

The reduced Gröbner basis of O_z is $\{Z_1^2 + Z_2^2 - r\}$, where $r = z_1^2 + z_2^2$. Hence $\mathbb{K}(z)^G = \mathbb{K}(r)$. The variety of $P = (Z_2)$ is a cross-section of degree 2 since the reduced Gröbner basis of I_z is $\{Z_2, Z_1^2 - r\}$.

Example 8.5 We consider the rational action of the additive group $\mathbb{K}$ on $\mathbb{K}^2$ given by:

$$\lambda \star z = \left(\frac{z_1}{1 + \lambda z_1}, \frac{z_2}{(1 + \lambda z_1)^2}\right).$$

Observe that $(\lambda + \mu) \star z = \lambda \star (\mu \star z)$ as prescribed by the group action axioms. Any point on the z_2-axis is a zero dimensional orbit whereas the generic orbits are one dimensional: They are the level sets of $\frac{z_2}{z_1^2}$. Concomitantly a reduced Gröbner basis of O_z is $\{Z_2 - \frac{z_2}{z_1^2} Z_1^2\}$. A generic line is a cross-section of degree 2. Yet the variety of $P = (Z_1 - 1)$ is a cross-section of degree 1: The Gröbner basis of I_z is $\{Z_1 - 1, Z_2 - \frac{z_2}{z_1^2}\}$. The rewriting described in Theorem 8.3 is then a simple replacement: $z_1 \mapsto 1$, $z_2 \mapsto r$, where $r = \frac{z_2}{z_1^2}$.

8.2.3 Replacement invariants

Cross-sections of degree 1 are of particular interest. The reduced Gröbner basis of ideal I_z is then $\{Z_1 - r_1, \ldots, Z_n - r_n\}$, for some $r_i \in \mathbb{K}(z)^G$. This was the case in Example 8.5. The rewriting of Theorem 8.3 is then a simple *replacement* of z_i by r_i: If R belongs to $\mathbb{K}(z)^G$ then $R(z_1, \ldots, z_n) = R(r_1(z), \ldots, r_n(z))$. Furthermore we know all the relationships on those invariants: A polynomial p that satisfies $p(r_1, \ldots, r_n) = 0$ belongs to P, the ideal of the cross-section.

This idea of an n-tuple of *replacement invariants* can be generalized for a cross-section of any degree if we allow algebraic invariants, i.e. algebraic functions of rational invariants. The reduced Gröbner basis of the ideal I_z considered in $\mathbb{K}(z)[Z]$ is also a reduced Gröbner basis of this ideal considered in $\mathbb{K}(z)^G[Z]$. If we have a section of degree e, the ideal I_z has e zeros in $\left(\overline{\mathbb{K}(z)^G}\right)^n$. Any such root $\xi = (\xi_1, \ldots, \xi_n)$ satisfies $r(z) = r(\xi)$ when $r \in \mathbb{K}(z)^G$. Furthermore if p is a polynomial such that $p(\xi_1, \cdots, \xi_n) = 0$ then $p \in P$, the ideal of the cross-section.

This fact is of course reminiscent of the property of the normalized invariants that appeared in the geometric construction of Section 8.2.1. And indeed, normalized invariants as defined in the neighborhood of a point on the cross-section where the action is well behaved, are zeros of I_z. This is illustrated by the case of rotations in Example 8.2 and then 8.4. This explains why normalized invariants end up being algebraic functions. Proper mathematical statements and further developments are to be found in [32].

The point we are making here is twofold. First, for rational actions of algebraic groups, which cover many situations, we have an algorithm to represent normalized invariants as algebraic functions. Second, we can work formally with normalized invariants as variables, as long as we subject them to the relationships defining the cross-section.

8.3 Differential invariants

The better known differential invariant is the curvature of a plane curve. It is an invariant for the Euclidean group, which consists of rotations and translations. If the curve is given as the graph $(x, u(x))$ of a smooth function $u : \mathbb{R} \to \mathbb{R}$, the curvature is $\sigma(x) = \sqrt{\frac{u_{xx}^2}{(1+u_x^2)^3}}$. It is an algebraic function of the derivatives of u. All other differential invariants can be obtained by differentiating the curvature with respect to arc length. This is an *invariant derivation* that can also be written explicitly in terms of the jet coordinates: $\mathcal{D} = (1 + u_x^2)^{-\frac{1}{2}} \frac{d}{dx}$.

If we now look at surfaces in $\mathbb{R}^3$ given as the graphs of a function $(x, y, u(x, y))$, two differential invariants come into play: Either the mean and Gauss curvatures or the principal curvatures. These pairs are not differentially independent of each other. Their derivatives, according to some invariant derivations, satisfy the Codazzi equation. Such a relationship is a *syzygy*.

In this section we describe algorithmic means to pinpoint invariant derivations, a generating set of differential invariants and their syzygies for a group action known solely by its infinitesimal generators. The content draws on [14, 29, 30].

8.3.1 Prolongations and invariant derivations

We consider a manifold $\mathcal{X} \times \mathcal{U}$, where $\mathcal{X}$ and $\mathcal{U}$ are open subset of $\mathbb{R}^m$ and $\mathbb{R}^n$, with respective coordinates $x = (x_1, \ldots, x_m)$ and $u = (u_1, \ldots, u_n)$. We are intrinsically looking locally at submanifolds of dimension m in an $m + n$ manifold: The submanifolds are given as the graphs of maps $u : \mathbb{R}^m \to \mathbb{R}^{m+n}$. For $\alpha = (\alpha_1, \ldots, \alpha_m) \in \mathbb{N}^m$, u_α stands for $\frac{\partial^{|\alpha|} u}{\partial x^\alpha} = \frac{\partial^{|\alpha|} u}{\partial x_1^{\alpha_1} \ldots \partial x_m^{\alpha_m}}$.

We discuss briefly the prolongation of an action on $\mathcal{X} \times \mathcal{U}$ to its jet space, in a simple coordinatized way. The coordinates for the k-th order jet space, noted $\mathrm{J}^k(\mathcal{X},\mathcal{U})$ or J^k for short, consist of x, u and u_α where $|\alpha| = \alpha_1 + \cdots + \alpha_m \leq k$. The functions on the infinite jet space J are subject to the *total derivations* $\mathrm{D}_1, \ldots, \mathrm{D}_m$ usually written

$$\mathrm{D}_i = \frac{\partial}{\partial x_i} + \sum_{u \in \mathcal{U},\, \alpha \in \mathbb{N}^m} u_{\alpha+\epsilon_i} \frac{\partial}{\partial u_\alpha}, \text{ for } 1 \leq i \leq m, \tag{8.1}$$

where ϵ_i stands for the m-tuple with 1 at the i-th position and 0 otherwise.

An action of a Lie group $\mathcal{G}$ on $\mathrm{J}^0 = \mathcal{X} \times \mathcal{U}$ can be prolonged in a unique geometrically meaningful way to an action $\mathcal{G} \times \mathrm{J}^k \to \mathrm{J}^k$. This prolongation can be made explicit as a change of variables in differential equations. Like many other operations in differential geometry and jet calculus, this can be performed with the Maple library *DifferentialGeometry* written by I. Anderson [2]. Similarly, the infinitesimal generators $\mathrm{v}_1, \ldots, \mathrm{v}_r$ of the action of $\mathcal{G}$ on $\mathcal{X} \times \mathcal{U}$ can be prolonged to vector fields $\mathrm{v}_1^k, \ldots, \mathrm{v}_r^k$ on J^k such that they are the infinitesimal generators for the prolonged action. A differential invariant of order k is then a local invariant of the action prolonged to J^k. We can use their infinitesimal characterization as a formal definition.

Definition 8.6 A differential invariant of order k is a function $f : \mathrm{J}^k \to \mathbb{R}$ such that $\mathrm{v}_1^k(f) = 0, \ldots, \mathrm{v}_r^k(f) = 0$.

For any k, a local cross-section to the orbits in J^k defines an invariantization and a set of normalized invariants:

$$\mathcal{I}^k = \{\bar{\iota}x_1, \ldots, \bar{\iota}x_m\} \cup \{\bar{\iota}u_\alpha \,|\, u \in \mathcal{U}, \, |\alpha| \leq k\}.$$

As discussed in Section 8.2.1, $\mathcal{I}^k$ forms a functionally generating set of local invariants on J^k and any differential invariant of order k and less can be written in terms of those by a simple rewriting.

As we prolong the action, the dimension of the orbits can only increase. It can not go beyond the dimension of the group. The stabilization order $\bar{s}$ is the order at which the maximal dimension of the orbits becomes stationary. With some mild assumptions on the group action on J^0, we can see that, for all $s \geq \bar{s}$, the orbits of the action on an open subset of J^s have the same dimension r as the group.

If we consider a local cross-section $\mathcal{P}$ on J^s, $s \geq \bar{s}$, given as the zero set of a map $p = (p_1, \ldots, p_r)$, the same equations define a local cross-section on J^{s+k}, for any $k > 0$. Thus, this determines the normalized invariants at any order $s + k$. The equations of the cross-section form a maximal independent set of relationships among them. This leaves us, however, with a rather infinite description of differential invariants.

A finite description of differential invariants can be obtained by introducing the concept of invariant derivations. Formally, these are total derivations that commute with the prolonged infinitesimal generators. When applied to a differential invariant, they give a differential invariant of a higher order.

The key fact is: When the stabilisation order is reached, and the orbits have therefore the same dimension as the group, the local cross-section defines a moving frame. With such a map in hand Fels and Olver [14] made explicit m invariant derivations $\mathcal{D} = (\mathcal{D}_1, \ldots, \mathcal{D}_m)$. Alternatively, [29, Theorem 3.4] can be consulted.

8.3.2 Generators

As discussed above the orbits of the action of $\mathcal{G}$ on J^s have the same dimension r as the group $\mathcal{G}$, when s is greater than the stabilization order. To a local cross-section on J^s we associated

- an invariantization process and thus normalized invariants of order s and greater
$$\mathcal{I}^{s+k} = \{\bar{\iota}x_1, \ldots, \bar{\iota}x_m\} \cup \{\bar{\iota}u_\alpha \,|\, u \in \mathcal{U},\ |\alpha| \leq s+k\}.$$
Any differential invariant can be written in terms of these by a simple substitution,
- a moving frame $\rho : J^s \to \mathcal{G}$ with which we can define invariant derivations $\mathcal{D} = (\mathcal{D}_1, \ldots, \mathcal{D}_m)$.

If applied to a differential invariant of order $s+k$, that can thus be written in terms of $\mathcal{I}^{s+k}$, the invariant derivations produce a differential invariant of order $s+k+1$. The important fact for a computational approach is that the action of the invariant derivations on the normalized invariants can be made explicit [14, Section 13] (alternatively [29, Theorem 3.6]).

Let the function $(p_1, \ldots, p_r)$ on J^s define the cross-section. Denote by $\mathrm{D}(P)$ the $m \times r$ matrix $(\mathrm{D}_i(p_j))_{i,j}$. The entries of $\mathrm{D}(P)$ are thus functions on J^{s+1}. Then $\mathrm{V}(P)$ is the $r \times r$ matrix $(\mathrm{v}_i(p_j))_{i,j}$. As $\mathcal{P}$ is transverse to the orbits of the action of $\mathcal{G}$ on J^s, the matrix $\mathrm{V}(P)$ has nonzero determinant along $\mathcal{P}$ and therefore in a neighborhood of each of its points.

Theorem 8.7 *Let K be the $m \times r$ matrix obtained by invariantizing of the entries of $\mathrm{D}(P)\,\mathrm{V}(P)^{-1}$. Then*
$$\mathcal{D}(\bar{\iota}f) = \bar{\iota}(\mathrm{D}f) - K\,\bar{\iota}(\mathrm{v}(f)).$$

The above theorem implies in particular the so called *recurrence formulae*:
$$\bar{\iota}(\mathrm{D}_i u_\alpha) = \mathcal{D}_i(\bar{\iota}u_\alpha) + \sum_{a=1}^{r} K_{ia}\,\bar{\iota}(\mathrm{v}_a(u_\alpha)),$$
where $K = \bar{\iota}(\mathrm{D}(P)\mathrm{V}(P)^{-1})$ has entries that are functions of $\mathcal{I}^{s+1}$. An inductive argument shows that any $\bar{\iota}u_\alpha$ can be written as a (rational) function of $\mathcal{I}^{s+1}$ and their derivatives. Combining this with the replacement property, we have a constructive way of rewriting any differential

invariant in terms of the elements of $\mathcal{I}^{s+1}$ and their derivatives. A differential invariant of order k is first trivially rewritten in terms of $\mathcal{I}^k$. If $k \leq s+1$ we are done. Otherwise, any element $\bar{\iota}u_\alpha$ of $\mathcal{I}^k$ with $|\alpha| = k$ is a $\bar{\iota}(\mathrm{D}_i u_\beta)$, for some $1 \leq i \leq m$ and $|\beta| = k-1$. We can thus write it as: $\bar{\iota}u_\alpha = \bar{\iota}(\mathrm{D}_i u_\beta) = \mathcal{D}_i(\bar{\iota}u_\beta) + \sum_{a=1}^{r} K_{ia}\, \bar{\iota}\,(\mathrm{v}_a(u_\beta))$. This involves only elements of $\mathcal{I}^{k-1}$ and their derivatives. Carrying on recursively we can rewrite everything in terms of the elements of $\mathcal{I}^{s+1}$ and their derivatives.

Theorem 8.8 *Any differential invariant of order $s+k$ can be written in terms of the elements of $\mathcal{I}^{s+1}$ and their invariant derivatives of order $k-1$ and less.*

A natural question is to determine a smaller set of differential invariants that is generating. A first result, initially stated in [49] in the case of coordinate cross-section and generalized in [29, Theorem 4.2], provides a generating set of at most $m(r+1)+n$ differential invariants if we impose some condition on the cross-section. This condition is actually quite natural: It has to define a cross-section at all orders.

Theorem 8.9 *If $P = (p_1, \ldots, p_r)$ defines a cross-section for the action on J such that $P_k = (p_1, \ldots, p_{r_k})$ defines a cross-section for the action on J^k, for all k, then $\mathcal{E} = \{\bar{\iota}(\mathrm{D}^i(p_j)) \,|\, 1 \leq i \leq m,\ 1 \leq j \leq r\}$ together with $\mathcal{I}^0$ form a generating set of differential invariants.*

The invariants in this generating set were named *edge invariants* in [49]. There is another set of invariants of the same size that was exhibited in [30] that does not require the cross-section to be of minimal order.

Theorem 8.10 *The union of 0-th order normalized invariants, $\mathcal{I}^0$, and the entries $\mathcal{K} = \{K_{ia} \,|\, 1 \leq i \leq m,\ 1 \leq a \leq r\}$, of the matrix $K = \bar{\iota}(\mathrm{D}(P)\mathrm{V}(P)^{-1})$, form a generating set of differential invariants.*

The generating property of these is actually a rather simple observation on the recurrence formula. These invariants carry an important geometric interpretation. They are the coefficients of the pull-back by the moving frame of the Maurer-Cartan forms on the group and are accordingly named the *Maurer-Cartan* invariants. Rewriting any normalized invariant, and hence any differential invariant, in terms of the Maurer-Cartan invariants and their invariant derivatives can be done by a simple recursive procedure.

8.3.3 Syzygies

The rewriting of any invariant in terms of the normalized invariants of order $s+1$, or the edge invariants, or the Maurer-Cartan invariants and their invariant derivatives is not unique. At each step of rewriting a $\bar{\iota}u_\alpha$, with $|\alpha| > s+1$, there might be several choices of pairs (i, β) such that $\beta + \epsilon_i = \alpha$ leading to rewriting $\bar{\iota}u_\alpha$ into $\mathcal{D}_i\left(\bar{\iota}u_\beta\right)$ with additional terms.

The first source of non-uniqueness comes from the fact that the invariant derivations do not commute. Yet their commutation rules are known explicitly [14, Section 13]. Those commutation rules can always be applied to rewrite any invariant derivatives in terms of monotone derivations $\mathcal{D}^\alpha = \mathcal{D}_1^{\alpha_1} \cdots \mathcal{D}_m^{\alpha_m}$.

Proposition 8.11 *For all* $1 \leq i, j \leq m$, $\left[\mathcal{D}_i, \mathcal{D}_j\right] = \sum_{k=1}^{m} \Lambda_{ijk} \, \mathcal{D}_k$ *where*

$$\Lambda_{ijk} = \sum_{c=1}^{r} K_{ic} \, \bar{\iota}(\mathrm{D}_j(\xi_{ck})) - K_{jc} \, \bar{\iota}(\mathrm{D}_i(\xi_{ck})),$$

$K = \bar{\iota}\left(D(P)\, V(P)^{-1}\right)$, *and* $\xi_{ck} = \mathrm{v}_c(x_k)$.

A *differential syzygy* is a relationship among a (generating) set of differential invariants and their monotone derivatives. A set of differential syzygies is complete if any other syzygy is inferred by those and their invariant derivatives.

The main point of [29] was to prove the completeness of the following set of syzygies for the normalized invariants of order $s+1$ in an appropriately formalized setting. The formalization also introduces some heavy notations that we skip to only render the essence. In the following theorem the $\bar{\iota}u_\alpha$ now stand for formal variables.

Theorem 8.12 *Consider a local cross-section given as the zero set of the functions* $p_1, \ldots, p_r$ *on* J^s, *where* s *is greater or equal to the stabilization order. A complete set of syzygies for* $\mathcal{I}^{s+1}$, *the normalized invariants of order* $s+1$, *is given by the union of the three following finite subsets:*

- $\mathfrak{R} = \left\{ p_1\left(\bar{\iota}x, \bar{\iota}u, \ldots, \bar{\iota}u^{(s)}\right) = 0, \ldots, p_r\left(\bar{\iota}x, \bar{\iota}u, \ldots, \bar{\iota}u^{(s)}\right) = 0 \right\}$
- $\mathfrak{S} = \left\{ S^i_{x_j} \mid 1 \leq i, j \leq m \right\} \cup \left\{ S^i_{u_\alpha} \mid |\alpha| \leq s, \, 1 \leq i \leq m \right\}$ *where*

$$S^i_{x_j} : \mathcal{D}_i(\bar{\iota}x_j) = \delta_{ij} - \sum_{a=1}^{r} K_{ia} \, \bar{\iota}(\mathrm{v}_a(x_j))$$

and

$$S^i_{u_\alpha} : \mathcal{D}_i(\bar{\iota}u_\alpha) = \bar{\iota}u_{\alpha+\epsilon_i} - \sum_{a=1}^{r} K_{ia}\,\bar{\iota}(\mathrm{v}_a(u_\alpha))$$

- $\mathfrak{T} = \{ T^i_{u_\beta} \mid |\beta| = s+1$ *and* $f_\beta < i \leq m \}$, *where* $f_\beta = \{\min j \mid \beta_j \neq 0\}$ *and* $T^i_{u_\beta}$ *is*

$$\mathcal{D}_i(\bar{\iota}u_\beta) - \mathcal{D}_{f_\beta}\left(\bar{\iota}u_{\beta+\epsilon_i-\epsilon_{f_\beta}}\right) = \sum_{a=1}^{r} K_{ia}\,\bar{\iota}(\mathrm{v}_a(u_{\beta+\epsilon_i-\epsilon_{f_\beta}})) - K_{f_\beta a}\,\bar{\iota}(\mathrm{v}_a(u_\beta)).$$

The elements of the first set $\mathfrak{R}$ are the relationships on the normalized invariants inherited from the equations of the cross-section. The second set $\mathfrak{S}$ consists of the recurrence formulae between the derivatives of normalized invariants of order s and less and the normalized invariants of order $s+1$. The syzygies $\mathfrak{T}$ consist of a subset of cross-derivatives for the invariants of order $s+1$. If $\alpha + \epsilon_i = \beta + \epsilon_j = \gamma$, then

$$\mathcal{D}_i(\bar{\iota}u_\alpha) = \bar{\iota}u_\gamma - \sum_{a=1}^{r} K_{ia}\bar{\iota}(\mathrm{v}_a(u_\alpha)), \quad \mathcal{D}_j(\bar{\iota}u_\beta) = \bar{\iota}u_\gamma - K_{ja}\bar{\iota}(\mathrm{v}_a(u_\beta)).$$

The difference thus forms a syzygy on $\mathcal{I}^{s+1}$. The set of all cross-derivatives is redundant since we can obtain some by simple combinations of the others.

The syzygies for the Maurer-Cartan invariants can also be spelt out. Due to their geometric meaning, their syzygies arise as the pullback, by the moving frame, of the structure equation on the group. The structure equations depend on some constants C_{abc} that describe the infinitesimal generators as a Lie algebra:

$$\mathrm{v}_a\mathrm{v}_b - \mathrm{v}_b\mathrm{v}_a = \sum_{c=1}^{r} C_{abc}\mathrm{v}_c.$$

Theorem 8.13 *Beside the relationships stemming from the equations of the cross-section, the Maurer-Cartan invariants, that are the entries of the matrix* $K = \bar{\iota}\left(\mathrm{D}(P)\mathrm{V}(P)^{-1}\right)$, *are subject to the syzygies*

$$\mathcal{D}_j(K_{ic}) - \mathcal{D}_i(K_{jc}) + \sum_{1\leq a<b\leq r} C_{abc}\,(K_{ia}K_{jb} - K_{ja}K_{ib}) + \sum_{k=1}^{m} \Lambda_{ijk}\,K_{kc} = 0,$$

for $1 \leq i < j \leq m$ *and* $1 \leq c \leq r$, *where* C_{abc} *are the structure constants and* Λ_{ijk} *are the coefficients of the commutation rules for the invariant derivations, written in terms of the Maurer-Cartan invariants.*

When it comes to the syzygies of the edge invariants, they can be obtained computationally from the syzygies on the normalized invariants by a differential elimination that takes place in the generalized differential algebra setting described in the next section. Differential elimination algorithms can be applied to further reduce the number of generating differential invariants, as was done in [33, 29]. Indeed, if a differential invariant can be written in terms of the others (and their invariant derivatives) it means that there is such a relationship in the differential ideal generated by the complete set of syzygies. This relationship can be exhibited by computing a representation of this differential ideal with an appropriate *elimination ranking*.

8.3.4 Example

Many simple examples, as well as computationally challenging ones, are treated in the papers [14, 29, 30] from which the material of this section is drawn as well as in, for instance, [33, 49, 39, 42, 40]. Here we wish to illustrate how the presented material is put into action on a well known example.

We consider the action of $SE(3)$ on surfaces in $\mathbb{R}^3$. We choose coordinate functions (x, y, u) for $\mathbb{R}^2 \times \mathbb{R}$, i.e. we consider x, y as the independent variables and u as the dependent variable. The infinitesimal generators of the classical action of the Euclidean group $SE(3)$ on $\mathbb{R}^3$ are:

$$\mathrm{v}_1 = \frac{\partial}{\partial x}, \qquad \mathrm{v}_2 = \frac{\partial}{\partial y}, \qquad \mathrm{v}_3 = \frac{\partial}{\partial v},$$
$$\mathrm{v}_4 = u\frac{\partial}{\partial y} - y\frac{\partial}{\partial u}, \quad \mathrm{v}_5 = x\frac{\partial}{\partial y} - y\frac{\partial}{\partial x}, \quad \mathrm{v}_6 = x\frac{\partial}{\partial u} - u\frac{\partial}{\partial x}.$$

The nonzero structure constants are given by the following commutators of the infinitesimal generators:

$$[\mathrm{v}_1, \mathrm{v}_5] = \mathrm{v}_2, \ [\mathrm{v}_1, \mathrm{v}_6] = \mathrm{v}_3, \ [\mathrm{v}_2, \mathrm{v}_4] = -\mathrm{v}_3, \ [\mathrm{v}_2, \mathrm{v}_5] = -\mathrm{v}_1, \ [\mathrm{v}_3, \mathrm{v}_4] = \mathrm{v}_2,$$
$$[\mathrm{v}_3, \mathrm{v}_6] = -\mathrm{v}_1, \ [\mathrm{v}_4, \mathrm{v}_5] = \mathrm{v}_6, \ [\mathrm{v}_4, \mathrm{v}_6] = -\mathrm{v}_5, \ [\mathrm{v}_5, \mathrm{v}_6] = \mathrm{v}_4.$$

Choose the classical cross-section defined by $P = (x, y, u, u_{10}, u_{01}, u_{11})$. The Maurer-Cartan matrix is then

$$K = \begin{pmatrix} 1 & 0 & 0 & 0 & \phi & \kappa \\ 0 & 1 & 0 & -\tau & \psi & 0 \end{pmatrix}$$

where

$$\kappa = \bar{\iota}u_{20}, \quad \tau = \bar{\iota}u_{02}, \quad \phi = \frac{\bar{\iota}u_{21}}{\bar{\iota}u_{20} - \bar{\iota}u_{02}}, \quad \text{and } \psi = \frac{\bar{\iota}u_{12}}{\bar{\iota}u_{20} - \bar{\iota}u_{02}}.$$

By Proposition 8.11 we have $[\mathcal{D}_2, \mathcal{D}_1] = \phi\, \mathcal{D}_1 + \psi\, \mathcal{D}_2$. The nonzero syzygies of Theorem 8.13 are:

$$\begin{aligned} \mathcal{D}_2(\kappa) - \phi\,(\kappa - \tau) &= 0, \\ \mathcal{D}_1(\tau) - \psi\,(\kappa - \tau) &= 0, \\ \mathcal{D}_2(\phi) - \mathcal{D}_1(\psi) - \kappa\,\tau - \phi^2 - \psi^2 &= 0. \end{aligned}$$

The first two syzygies imply that

$$\phi = \frac{\mathcal{D}_2(\kappa)}{\kappa - \tau}, \quad \psi = \frac{\mathcal{D}_1(\tau)}{\kappa - \tau}.$$

It follows that $\{\kappa, \tau\}$ form a generating set. From their analytic expressions [21] we can write the Gauss and mean curvatures in terms of the normalized invariants and eventually in terms of the Maurer-Cartan invariants:

$$\begin{aligned} \sigma &= \frac{u_{20}u_{02} - u_{11}}{(1 + u_{10}^2 + u_{01}^2)^2} = \bar{\iota}u_{20}\,\bar{\iota}u_{02} = \kappa\,\tau, \\ \pi &= \frac{1}{2}\,\frac{(1 + u_{01}^2)u_{20} - 2_{10}u_{01}u_{11} + (1 + u_{10}^2)u_{20}}{(1 + u_{10}^2 + u_{01}^2)^{\frac{3}{2}}} \\ &= \tfrac{1}{2}(\bar{\iota}u_{20} + \bar{\iota}u_{02}) = \tfrac{1}{2}(\kappa + \tau). \end{aligned}$$

Our generating invariants $\{\kappa, \tau\}$ are thus the principal curvatures. Substituting ϕ and ψ in the last syzygy we retrieve the Codazzi equation:

$$\mathcal{D}_2\left(\frac{\mathcal{D}_2(\kappa)}{\kappa - \tau}\right) - \mathcal{D}_1\left(\frac{\mathcal{D}_1(\tau)}{\kappa - \tau}\right) = \left(\frac{\mathcal{D}_2(\kappa)}{\kappa - \tau}\right)^2 + \left(\frac{\mathcal{D}_1(\tau)}{\kappa - \tau}\right)^2 + \kappa\tau.$$

8.4 Generalized differential algebras

What we have so far is a characterization of a generating set for the differential invariants of the action of a Lie group, a way to rewrite any differential invariant in terms of their monotone derivatives and a complete set of the differential relationships they satisfy. In the case of the rational action of an algebraic group we also have an algorithm to compute explicitly those generating differential invariants. Importantly though, the complete set of syzygies and the rewriting can actually be obtained independently of the explicit expression of the generating set. We can thus work formally with those.

The set of differential invariants is adequately represented by a differential algebra that is the quotient of a differential polynomial ring by a differential ideal. The differential polynomial ring has a differential indeterminate standing for each generating differential invariant. The syzygies live in this differential polynomial ring. They generate the differential ideal to quotient it with.

After the advent of Gröbner bases as a fundamental tool in computational algebra, efforts were made to provide practical algorithms for differential elimination and completion based on the concepts of differential algebra stemming out of work of J. Ritt and E. Kolchin [57, 36]. Complete lecture notes [25, 26] are available for a detailed understanding of the concepts involved and the algorithms underlying the *diffalg* library in Maple. A representative sample of the original articles is [6, 7, 24, 38, 41, 55, 56]. All these address systems of classical differential equations, i.e. work with commuting derivations. Dealing with differential invariants and invariant derivations brought a new algorithmic challenge, first discussed in [39].

In the *algebra of differential invariants*, the derivations do not commute. Furthermore, the commutation coefficients involve the differential indeterminates themselves, or even their derivatives. An adequate concept of differential polynomial ring was provided in [28]. The rather amazing part is that, in the end, it differs very little from the classical case. In particular, all the algorithms that perform differential elimination and completion can be extended to this new context by just changing the way derivations act on derivatives.

Let $\mathcal{Y} = \{y_1, \ldots, y_n\}$ be a set of differential indeterminates. We define $\mathbb{K}[\![\mathcal{Y}]\!] = \mathbb{K}[\![y_1, \ldots, y_n]\!]$ to be the polynomial ring in the infinitely many variables $\{y_\alpha \mid y \in \mathcal{Y},\ \alpha \in \mathbb{N}^m\}$, called the derivatives. $\mathbb{K}$ is a field of characteristic zero to which the derivations can be restricted. In the classical case it is a field of functions in the independent variables. For the algebra of differential invariants, it is a field of constants. We endow $\mathbb{K}[\![\mathcal{Y}]\!]$ with a set of m derivations $\mathcal{D} = \{\mathcal{D}_1, \ldots, \mathcal{D}_m\}$ defined recursively on $\{y_\alpha\}$ by

$$\mathcal{D}_i(y_\alpha) = \begin{cases} y_{\alpha+\epsilon_i} & \text{if } \alpha_1 = \cdots = \alpha_{i-1} = 0 \\ & \\ \mathcal{D}_j \mathcal{D}_i\left(y_{\alpha-\epsilon_j}\right) & \text{where } j < i \text{ is s.t. } \alpha_j > 0 \\ \quad + \sum_{l=1}^m \Lambda_{ijl}\, \mathcal{D}_l\left(y_{\alpha-\epsilon_j}\right) & \text{while } \alpha_1 = \cdots = \alpha_{j-1} = 0 \end{cases}$$

where the family $\{\Lambda_{ijl}\}_{1 \le i,j,l \le n}$ of elements of $\mathbb{K}[\![\mathcal{Y}]\!]$ is such that:

- $\Lambda_{ijl} = -\Lambda_{jil}$
- $\displaystyle\sum_{\mu=1}^{m} \Lambda_{ij\mu}\Lambda_{\mu kl} + \Lambda_{jk\mu}\Lambda_{\mu il} + \Lambda_{ki\mu}\Lambda_{\mu jl} = \mathcal{D}_k(\Lambda_{ijl}) + \mathcal{D}_i(\Lambda_{jkl}) + \mathcal{D}_j(\Lambda_{kil})$.

This latter relationship stands for a Jacobi identity and is quite natural. It insures that the $\{y_\alpha\}$ remain algebraically independent. It is certainly satisfied by the commutation coefficients that arise for the invariant derivations in Section 8.3.

Definition 8.14 An *admissible ranking* on $\mathbb{K}[\![\mathcal{Y}]\!]$ is a total order $\prec$ on the set of derivatives $\{y_\alpha\}$ such that:

- $|\alpha| < |\beta| \ \Rightarrow\ y_\alpha \prec y_\beta$, $\forall \alpha, \beta \in \mathbb{N}^m$, $\forall y \in \mathcal{Y}$;
- $y_\alpha \prec z_\beta \ \Rightarrow\ y_{\alpha+\gamma} \prec z_{\beta+\gamma}$, $\forall \alpha, \beta, \gamma \in \mathbb{N}^m$, $\forall y, z \in \mathcal{Y}$;
- $\sum_{l=1}^{m} \Lambda_{ijl} \mathcal{D}_l(y_\alpha) \prec y_{\alpha+\epsilon_i+\epsilon_j}$, for all $1 \leq i, j \leq m$, $\forall y \in \mathcal{Y}$.

If $\mathbb{K}[\![\mathcal{Y}]\!]$ can be endowed with an admissible ranking then it is proved in [28] that:

- $\mathcal{D}_i\mathcal{D}_j(p) - \mathcal{D}_j\mathcal{D}_i(p) = \sum_{l=1}^{m} \Lambda_{ijl}\mathcal{D}_l(p), \quad \forall p \in \mathbb{K}[\![\mathcal{Y}]\!]$,
- $\mathcal{D}^\beta(y_\alpha) - y_{\alpha+\beta}$ ranks lower than $y_{\alpha+\beta}$.

In this case we shall say that $\mathbb{K}[\![\mathcal{Y}]\!]$ is a *differential polynomial ring with non-trivial commutation rules for the derivations* $\{\mathcal{D}_1, \ldots, \mathcal{D}_m\}$.

Let us make a couple of remarks. If the Λ_{ijl} are differential polynomials that involve derivatives of order one or less then any orderly ranking is admissible. This is automatically the case for the algebra of differential invariants when normalized or Maurer-Cartan invariants are selected as generators. Elimination rankings can also be admissible. If the coefficients Λ_{ijl} involve only derivatives of a subset $\mathcal{Z} \subset \mathcal{Y}$ of the differential indeterminates then we can consider a ranking that eliminates $\mathcal{Y} \setminus \mathcal{Z}$.

In classical differential algebra, i.e. when the derivations commute, rankings are only subject to the conditions:

- $y_\alpha \prec y_{\alpha+\gamma}$, $\forall y \in \mathcal{Y}$, $\alpha, \gamma \in \mathbb{N}^m$
- $y_\alpha \prec z_\beta \ \Rightarrow\ y_{\alpha+\gamma} \prec z_{\beta+\gamma}$, $\forall \alpha, \beta, \gamma \in \mathbb{N}^m$, $\forall y, z \in \mathcal{Y}$.

There we can consider rankings that are not semi-orderly as the one given by:

$$y_\alpha \prec y_\beta \ \Leftrightarrow\ \exists i \text{ such that } \alpha_1 = \beta_1, \ldots, \alpha_{i-1} = \beta_{i-1} \text{ and } \alpha_i > \beta_i.$$

This type of ranking favors one of the derivations and cannot be handled in a context where derivations do not commute.

As mentioned above, the concepts and main results of classical differential algebra go through with nearly identical proofs. In particular any radical differential ideal is finitely generated and has a unique minimal decomposition into prime differential ideals. The algorithms also work, with the restriction of ranking described above. The Maple library *diffalg* was extended to handle this new type of differential polynomial ring [5, 27]. The concepts and algorithms underlying it are described in great detail in [25, 26].

8.5 Prospects

In Section 8.2 we pointed out how obtaining a cross-section of degree 1 was interesting. Yet the question remains of when such a cross-section exists, and how to compute one. When no such cross-section can be found, one must still address the issue of determining the relationships on the generating rational invariants we obtained as the coefficients of the reduced Gröbner basis. We can always resort to another algebraic elimination to find them, but some geometric understanding should help the task.

In Section 8.3 we exhibited different sets of generating invariants. An open question is what is the minimal cardinality of such a set. For curves in affine n-space it is known that n differential invariants suffice as generators [19, 47]. One question I cannot answer though is how we can pinpoint n such generators from the Maurer-Cartan invariants. For surfaces in 3-space, under the Euclidean and affine group, two curvatures with a syzygy describe the algebra of differential invariants. Using the tools presented here and their implementation, it was shown that the same is true for the projective and conformal group acting on surfaces [33]. Some cases of 3-dimensional submanifolds in 4-space were also given a computational treatment in [29, Section 7]. For the general question of what is a minimal set of generating differential invariants, further theoretical development would be helpful.

The seminal article [14] offered methodological tools for deciding the equivalence of manifolds, the long standing problem attached to Cartan's moving frame method. My interest in the present subject stemmed from the demonstration in [39] of how those tools could be used for solving a differential elimination problem through symmetry reduction. An example of how I think symmetry reduction, with a view towards differential elimination, should be tackled is presented as motivation in [28]. While some of the theoretical aspects have been worked out in the mean time, this particular problem has resisted my computational attempts at solving it. There is a need for more computational success on this idea so as to validate this approach.

A natural way to further advance the topic has been initiated in [52, 50, 51]. A methodological theory for pseudo-groups is introduced there in the same way as [14] started the topic for finite dimensional Lie groups.

References

[1] W. Adams and P. Loustaunau, *An Introduction to Gröbner Bases.* AMS, Providence, 1994.

[2] I. M. Anderson, Maple packages and Java applets for classification problems in geometry and algebra. In *Foundations of Computational Mathematics: Minneapolis, 2002,* London Math. Soc. Lecture Note Ser., **312**, 193–206, Cambridge Univ. Press, Cambridge, 2004.

[3] T. Becker and V. Weispfenning, *Gröbner Bases - A Computational Approach to Commutative Algebra.* Springer-Verlag, New York, 1993.

[4] M. Berger and B. Gostiaux, *Differential Geometry: Manifolds, Curves, and Surfaces.* Springer-Verlag, New York, 1988.

[5] F. Boulier and E. Hubert, DIFFALG*: description, help pages and examples of use.* Symbolic Computation Group, University of Waterloo, Ontario, Canada, 1998.

[6] F. Boulier, D. Lazard, F. Ollivier, and M. Petitot, Representation for the radical of a finitely generated differential ideal. In A. H. M. Levelt, editor, *ISSAC'95.* ACM Press, New York, 1995.

[7] F. Boulier, D. Lazard, F. Ollivier, and M. Petitot, Computing representations for radicals of finitely generated differential ideals. *Appl. Algebra Engrg. Comm. Comput.*, **20**, (2009), 73–121.

[8] M. Boutin and G. Kemper, On reconstructing configurations of points in $\mathbb{P}^2$ from a joint distribution of invariants. *Appl. Algebra Engrg. Comm. Comput.*, **15**, (2005), 361–391.

[9] P. Chossat and R. Lauterbach, *Methods in Equivariant Bifurcations and Dynamical Systems*, volume 15 of *Advanced Series in Nonlinear Dynamics.* World Scientific Publishing Co. Inc., River Edge, NJ, 2000.

[10] K.-S. Chou and C.-Z. Qu, Integrable equations arising from motions of plane curves. II. *J. Nonlinear Sci.*, **13**, (2003), 487–517.

[11] P. Comon and B. Mourrain, Decomposition of quantics in sums of power of linear forms. *Signal Processing*, **52**, (1996), 96–107.

[12] D. Cox, J. Little, and D. O'Shea, *Ideals, Varieties, and Algorithms.* Springer-Verlag, 1992.

[13] H. Derksen and G. Kemper, *Computational Invariant Theory.* Invariant Theory and Algebraic Transformation Groups I. Springer-Verlag, Berlin, 2002.

[14] M. Fels and P. J. Olver, Moving coframes. II. Regularization and theoretical foundations. *Acta Appl. Math.*, **55**, (1999), 127–208.

[15] J. Flusser, T. Suk, and B. Zitová, *Moments and Moment Invariants in Pattern Recognition.* Wiley and Sons Ltd., 2009.

[16] R. B. Gardner, *The Method of Equivalence and its Applications.* CBMS-NSF Regional Conference Series in Applied Mathematics, **58**, SIAM, Philadelphia, 1989.

[17] K. Gatermann, *Computer Algebra Methods for Equivariant Dynamical Systems.* Lecture Notes in Mathematics, **1728**, Springer-Verlag, Berlin, 2000.

[18] J. H. Grace and A. Young, *The Algebra of Invariants.* Cambridge Univ. Press, Cambridge, 1903.

[19] M. L. Green, The moving frame, differential invariants and rigidity theorems for curves in homogeneous spaces. *Duke Math. Journal*, **45**, (1978), 735–779.

[20] G-M. Greuel and G. Pfister, *A* **Singular** *Introduction to Commutative Algebra.* Springer-Verlag, Berlin, 2002.

[21] H. W. Guggenheimer, *Differential Geometry.* McGraw-Hill Book Co., Inc., New York, 1963.

[22] G. Gurevich, *Foundations of the Theory of Algebraic Invariants.* Noordhoff, 1964.

[23] D. Hilbert, *Theory of Algebraic Invariants.* Cambridge University Press, Cambridge, 1993.

[24] E. Hubert, Factorisation free decomposition algorithms in differential algebra. *J. Symbolic Comp.*, **29**, (2000), 641–662.

[25] E. Hubert, Notes on triangular sets and triangulation-decomposition algorithms I: Polynomial systems. In *Symbolic and Numerical Scientific Computing*, F. Winkler and U. Langer, editors, in Lecture Notes in Computer Science, **2630**, 1–39, Springer-Verlag, Heidelberg, 2003.

[26] E. Hubert, Notes on triangular sets and triangulation-decomposition algorithms II: Differential systems. In *Symbolic and Numerical Scientific Computing*, F. Winkler and U. Langer, editors, in Lecture Notes in Computer Science, **2630**, 40–87, Springer-Verlag, Heidelberg, 2003.

[27] E. Hubert, DIFFALG*: extension to non commuting derivations.* INRIA, Sophia Antipolis, 2005.

[28] E. Hubert, Differential algebra for derivations with nontrivial commutation rules. *J. Pure Applied Algebra*, **200**, (2005), 163–190.

[29] E. Hubert, Differential invariants of a Lie group action: syzygies on a generating set. *J. Symbolic Comp.*, **44**, (2009), 382–416.

[30] E. Hubert, Generation properties of Maurer-Cartan invariants. Preprint `http://hal.inria.fr/inria-00194528`, 2012.

[31] E. Hubert and I. A. Kogan, Rational invariants of a group action. Construction and rewriting. *J. Symbolic Comp.*, **42**, (2007), 203–217.

[32] E. Hubert and I. A. Kogan, Smooth and algebraic invariants of a group action. Local and global constructions. *Foundations of Comp. Math.*, **7**, (2007), 455–493.

[33] E. Hubert and P. J. Olver, Differential invariants of conformal and projective surfaces. *Symmetry Integrability and Geometry: Methods and Applications*, 3(097), (2007).

[34] G. Jensen, *Higher Order Contact of Submanifolds of Homogeneous Spaces.* Springer-Verlag, Berlin, 1977.

[35] G. Kemper, The computation of invariant fields and a new proof of a theorem by Rosenlicht. *Transformation Groups*, **12**, (2007), 657–670.

[36] E. R. Kolchin, *Differential Algebra and Algebraic Groups.* Academic Press, New York, 1973.

[37] M. Kreuzer and L. Robbiano, *Computational Commutative Algebra. 1.* Springer-Verlag, Berlin, 2000.

[38] E. L. Mansfield, *Differential Gröbner Bases.* PhD thesis, University of Sydney, 1991.

[39] E. L. Mansfield, Algorithms for symmetric differential systems. *Foundations of Comp. Math.*, **1**, (2001), 335–383.

[40] E. L. Mansfield, *A Practical Guide to the Invariant Calculus.* Cambridge University Press, 2010.

[41] E. L. Mansfield and P. A. Clarkson, Application of the differential algebra package diffgrob2 to classical symmetries of differential equations. *J. Symbolic Comp.*, **23**, (1997), 517–533.

[42] E. L. Mansfield and P. H. van der Kamp, Evolution of curvature invariants and lifting integrability. *J. Geom. Phys.*, **56**, (2006), 1294–1325.

[43] G. Marí Beffa, Projective-type differential invariants and geometric curve evolutions of KdV-type in flat homogeneous manifolds. *Annales de l'Institut Fourier (Grenoble)*, **58**, (2008), 1295–1335.

[44] J. Müller-Quade and T. Beth, Calculating generators for invariant fields of linear algebraic groups. In *Applied Algebra, Algebraic Algorithms and Error-Correcting Codes (Honolulu, HI, 1999)*, of Lecture Notes in Computer Science, **1719**, 392–403, Springer, Berlin, 1999.

[45] J. L. Mundy, A. Zisserman, and D. A. Forsyth, editors, *Applications of Invariance in Computer Vision, Second Joint European - US Workshop, Ponta Delgada, Azores, Portugal, October 9-14, 1993, Proceedings*, **825**, Springer, 1994.

[46] J. L. Mundy and A. Zisserman, editors, *Geometric Invariance in Computer Vision.* MIT Press, 1992.

[47] P. J. Olver, *Applications of Lie Groups to Differential Equations.* Springer-Verlag, New York, 1986.

[48] P. J. Olver, *Equivalence, Invariants and Symmetry.* Cambridge University Press, 1995.

[49] P. J. Olver, Generating differential invariants. *J. Math. Anal. Appl.*, **333**, (2007), 450–471.

[50] P. J. Olver and J. Pohjanpelto, Moving frames for Lie pseudo-groups. *Canad. J. Math.*, **60**, (2008), 1336–1386.

[51] P. J. Olver and J. Pohjanpelto, Differential invariant algebras of Lie pseudo-groups. *Adv. Math.*, **222**, (2009), 1746–1792.

[52] P. J. Olver and J. Pohjanpelto, Maurer-Cartan forms and the structure of Lie pseudo-groups. *Selecta Mathematica*, **11**, (2005), 99–126.

[53] L. V. Ovsiannikov, *Group Analysis of Differential Equations.* Academic Press Inc., New York, 1982.

[54] V. L. Popov and E. B. Vinberg, Invariant theory. In *Algebraic Geometry. IV*, number 55 in Encyclopaedia of Mathematical Sciences, 122–278. Springer-Verlag, 1994.

[55] G. J. Reid, Algorithms for reducing a system of pde to standard form, determining the dimension of its solution space and calculating its taylor series solution. *European J. Applied Math.*, **2**, (1991), 293–318.

[56] G. J. Reid and A. Boulton, Reduction of differential equations to standard form and their integration using directed graphs. In *ISSAC'91*, S. M. Watt, editor, 308–312. ACM Press, 1991.

[57] J. F. Ritt, *Differential Algebra.* volume XXXIII of *Colloquium Publications*, AMS, 1950.

[58] M. Rosenlicht, Some basic theorems on algebraic groups. *Amer. J. Math.*, **78**, (1956), 401–443.

[59] C. Shakiban and P. Lloyd, Signature curves statistics of DNA supercoils. In *Geometry, Integrability and Quantization*, 203–210, Softex, Sofia, 2004.

[60] B. Sturmfels, *Algorithms in Invariant Theory.* Springer-Verlag, Vienna, 1993.

9

Through the Kaleidoscope: Symmetries, Groups and Chebyshev-Approximations from a Computational Point of View

Hans Z. Munthe-Kaas, Morten Nome and Brett N. Ryland

Department of Mathematics, University of Bergen

Abstract

In this paper we survey parts of group theory, with emphasis on structures that are important in design and analysis of numerical algorithms and in software design. In particular, we provide an extensive introduction to Fourier analysis on locally compact abelian groups, and point towards applications of this theory in computational mathematics. Fourier analysis on non-commutative groups, with applications, is discussed more briefly. In the final part of the paper we provide an introduction to multivariate Chebyshev polynomials. These are constructed by a kaleidoscope of mirrors acting upon an abelian group, and have recently been applied in numerical Clenshaw–Curtis type numerical quadrature and in spectral element solution of partial differential equations, based on triangular and simplicial subdivisions of the domain.

9.1 Introduction

Group theory is the mathematical language of symmetry. As a mature branch of mathematics, with roots going almost two centuries back, it has evolved into a highly technical discipline. Many texts on group theory and representation theory are not readily accessible to applied mathematicians and computational scientists, and the relevance of group theoretical techniques in computational mathematics is not widely recognized.

Nevertheless, it is our conviction that knowledge of central parts of group theory and harmonic analysis on groups is invaluable also for computational scientists, both as a language to unify, analyze and generalize computational algorithms and also as an organizing principle of mathematical software construction.

The first part of the paper provides a quite detailed self-contained introduction to Fourier analysis in the language of compact abelian groups. Abelian groups are fundamental structures of major importance in computations as they describe spaces with commutative shift operations. Both continuous and discrete abelian groups are omnipresent as computational domains in applied mathematics. Classical Fourier analysis can be understood as the theory of linear operators which commute with translations in space; examples being differential operators with constant coefficients, and other linear operators which are invariant under change of time or space. A central theme in computation is the interplay between continuous and discrete structures, between continuous mathematical models and their discretizations. Many aspects of discretizations can be understood via subgroups and sub-lattices of continuous abelian groups. The duality between time/space and frequency/wave numbers provided by the Fourier transform is another central topic which mathematically is expressed via Pontryagin duality of abelian groups.

Symmetries is a core topic of all group theory. Translation invariance of linear operators is one type of symmetry. Another kind of symmetry is invariance of operators under isometries acting upon the domain, e.g. reflection symmetries. In the second part of this paper we will discuss the importance of certain kaleidoscopic groups generated by a set of mirrors acting upon a domain. We will review basic properties of multivariate versions of Chebyshev polynomials, which are constructed by the folding of exponential functions under the action of kaleidoscopes. Theoretically these have remarkable properties, both from an approximation theoretical point of view and also with respect to computational complexity. However, until recently they have not been applied in computations, probably due to the quite complicated underlying theory. We will briefly review recent work where multivariate Chebyshev polynomials are applied in numerical quadrature and in spectral element solution of PDEs, and discuss some advantages and difficulties in this line of work.

9.2 Fourier analysis on groups

Classical Fourier analysis is intimately connected with the theory of locally compact abelian groups. In mathematical analysis this view allows a unified presentation of the Fourier transform, Fourier series and the Discrete Fourier Transform [31]. In computational science, some understanding of the group structures underlying discrete and continuous Fourier analysis is invaluable both as a language to discuss computational aspects of Fourier transforms and sampling theory, as well as an organizing principle of mathematical software [14]. In this section we provide a self-contained review of Fourier analysis on abelian groups. We conclude this section with a brief discussion of generalized Fourier transforms on non-abelian groups and point to some applications of representation theory in computational science. We refer to [5, 7, 11, 15, 16, 27, 31, 33] for more details on the topics of this chapter.

9.2.1 Locally compact abelian groups

A Locally Compact Abelian group (LCA) is a locally compact topological space G which is also an abelian group. Thus G has an identity element 0, a group operation $+$ and a unary $-$ operation, such that $a+b=b+a$, $a+(b+c)=(a+b)+c$ and $a+(-a)=0$ for all $a,b,c\in G$. For our discussion it suffices to consider the *elementary* LCAs and groups isomorphic to one of these:

Definition 9.1 The *elementary LCAs* are:

- The *reals* $\mathbb{R}$ under addition, with the standard definition of open sets.
- The *integers* $\mathbb{Z}$ under addition (with the discrete topology). This is also known as the infinite cyclic group.
- The 1-dimensional *torus*, or *circle* $T=\mathbb{R}/2\pi\mathbb{Z}$ defined as $[0,2\pi)\subset\mathbb{R}$ under addition modulo 2π, with the circle topology.
- The *cyclic group* of order k, $\mathbb{Z}_k=\mathbb{Z}/k\mathbb{Z}$, which consists of the integers $0,1,\ldots,k-1$ under addition modulo k (with the discrete topology).
- *Direct products* of the above spaces, $G\times H$, in particular $\mathbb{R}^n$ (real n-space), T^n (the n-torus) and all finitely generated abelian groups.

It is often convenient to consider these groups in their isomorphic *multiplicative form*. E.g. the multiplicative group of complex numbers of modulus 1, denoted $\mathbb{T}=\{\,z\in\mathbb{C}\mid |z|=1\,\}$ is isomorphic to T via the exponential map $T\ni\theta\mapsto\exp(i\theta)\in\mathbb{T}$. More generally, if G is an additive abelian group with elements λ, we denote a corresponding multiplicative group as e^G with elements e^λ, multiplicative group operation $e^\lambda e^{\lambda'}=$

$e^{\lambda+\lambda'}$ and unit $1 = e^0$. The exponential notation is here just a formal change of notation, a 'syntactic sugar' to simplify certain calculations.

An LCA with a finite subset of generators is called a *Finitely Generated Abelian group* (FGA). These, and in particular the finite ones, are the basic domains of computational Fourier analysis and the Fast Fourier Transform. A complete classification of these is simple and very useful:

Theorem 9.2 (Classification of FGA) *If G is an FGA, then G is isomorphic to a group of the form*

$$\mathbb{Z}^n \times \mathbb{Z}_{(p_0)^{n_0}} \times \mathbb{Z}_{(p_1)^{n_1}} \times \cdots \times \mathbb{Z}_{(p_{k-1})^{n_{k-1}}}$$

where p_i are primes, $p_0 \leq p_1 \leq \cdots \leq p_{n-1}$ and $n_i \leq n_{i+1}$ whenever $p_i = p_{i+1}$. Furthermore, G is also isomorphic to a group of the form

$$\mathbb{Z}^n \times \mathbb{Z}_{m_0} \times \mathbb{Z}_{m_1} \times \cdots \times \mathbb{Z}_{m_{\ell-1}}$$

where m_i divides m_{i+1}. In both forms the representation is unique, i.e. two FGA are isomorphic if and only if they can be transformed into the same canonical form[1].

9.2.2 Dual groups and the Fourier transform

Group homomorphisms are fundamental in describing sampling theory, fast Fourier transforms and convolution theorems.

Definition 9.3 For two LCA H and G, we let $\hom(H, G)$ denote the set of continuous homomorphisms from H to G, i.e. all continuous maps $\phi\colon H \to G$ such that

$$\phi(h + h') = \phi(h) + \phi(h') \quad \text{for all } h, h' \in H.$$

In particular $\chi \in \hom(G, \mathbb{T})$ are homomorphisms satisfying $\chi(x+x') = \chi(x)\chi(x')$ for all $x \in G$, where $\chi(x) = e^{i\theta(x)}$ and $\theta(x) \in \mathbb{R}$. These functions χ are called the *characters* of G. The characters are always given by exponential maps and form the basis for the Fourier transform. For two characters χ, χ' it is clear that the product $(\chi\chi')(g) = \chi(g)\chi'(g)$ is also a character, and also the complex conjugate $\overline{\chi}$ is a character, which is the multiplicative inverse of χ. Thus $\hom(G, \mathbb{T})$ is an abelian group called the *dual group* of G, denoted $\widehat{G}$. With the compact-open topology $\widehat{G}$ is also an LCA, thus we can define the double dual $\widehat{\widehat{G}}$. Pontryagin's

[1] The isomorphisms are provided by the *Chinese remainder theorem*.

duality theorem states that G and $\widehat{\widehat{G}}$ are isomorphic LCAs. Due to this symmetry, we adopt a slightly different definition of dual LCAs. In the following, we consider both G and $\widehat{G}$ as additive abelian groups and define the duality between these via a dual pairing:

Definition 9.4 (Dual pair) Two LCAs G and $\widehat{G}$ are called a *dual pair of LCAs* if there exists a continuous function $\langle\cdot,\cdot\rangle : \widehat{G}\times G \to \mathbb{T}$ such that

$$\begin{aligned} \hom(G,\mathbb{T}) &= \left\{ \langle k,\cdot\rangle \mid k \in \widehat{G} \right\} \\ \hom(\widehat{G},\mathbb{T}) &= \{ \langle \cdot, x\rangle \mid x \in G \}. \end{aligned}$$

The definition implies that

$$\begin{aligned} \langle k+k', x\rangle &= \langle k,x\rangle\cdot\langle k',x\rangle \\ \langle k, x+x'\rangle &= \langle k,x\rangle\cdot\langle k,x'\rangle \\ \langle 0,x\rangle &= \langle k,0\rangle = 1 \\ \overline{\langle k,x\rangle} &= \langle -k,x\rangle = \langle k,-x\rangle. \end{aligned}$$

Furthermore, if $\langle k,x\rangle = 1$ for all k, then $x = 0$, and if $\langle k,x\rangle = 1$ for all x, then $k = 0$. If we work with pairings on different groups G, H, we write $\langle\cdot,\cdot\rangle_G$ and $\langle\cdot,\cdot\rangle_H$ to distinguish them.

Every LCA has a translation invariant Haar measure, yielding an integral $\int_G f(x)d\mu$ which is uniquely defined up to a scaling. For $\mathbb{R}^n$ and T^n, this is the standard integral, and for the discrete $\mathbb{Z}_k$ and $\mathbb{Z}$, it is the sum over the elements. The characters are orthogonal under the inner product defined by the Haar measure. Here $L^2(G)$ denotes square integrable functions on G.

Definition 9.5 (Fourier transform) The Fourier transform is a unitary[2] map $\widehat{\ }: L^2(G) \to L^2(\widehat{G})$ given as

$$\widehat{f}(k) = \int_G \langle -k,x\rangle f(x)dx.$$

There exists a constant C such that the inverse transform is given as

$$f(x) = \frac{1}{C}\int_{\widehat{G}} \langle k,x\rangle \widehat{f}(k)dk.$$

Sometimes we will write $\mathcal{F}(f)$ or $\mathcal{F}_G(f)$ instead of $\widehat{f}$.

Note that $\widehat{G}$ is discrete if and only if G is compact. In this case the

[2] Unitary up to a scaling.

inversion formula becomes a sum, and $C = \int_G 1dx = \text{vol}(G)$. The following table presents the dual pairs for the elementary groups $\mathbb{R}$, T, $\mathbb{Z}$ and $\mathbb{Z}_n$.

G	$\widehat{G}$	$\langle\cdot,\cdot\rangle$	$\widehat{f}(\cdot)$	$f(\cdot)$
$x \in \mathbb{R}$	$\omega \in \mathbb{R}$	$e^{2\pi i\omega x}$	$\int_{-\infty}^{\infty} e^{-2\pi i\omega x} f(x)dx$	$\int_{-\infty}^{\infty} e^{2\pi i\omega x} \widehat{f}(\omega)d\omega$
$x \in T$	$k \in \mathbb{Z}$	$e^{2\pi ikx}$	$\int_0^1 e^{-2\pi ikx} f(x)dx$	$\sum_{k=-\infty}^{\infty} e^{2\pi ikx} \widehat{f}(k)$
$j \in \mathbb{Z}_n$	$k \in \mathbb{Z}_n$	$e^{\frac{2\pi ikj}{n}}$	$\sum_{j=0}^{n-1} e^{\frac{-2\pi ikj}{n}} f(j)$	$\frac{1}{n}\sum_{k=0}^{n-1} e^{\frac{2\pi ikj}{n}} \widehat{f}(k)$

Multidimensional versions are given by the componentwise formulae:

$$\begin{aligned}
x &= (x_1, x_2) \in G = G_1 \times G_2 \\
k &= (k_1, k_2) \in \widehat{G} = \widehat{G_1} \times \widehat{G_2} \\
\langle k, x\rangle &= \langle k_1, x_1\rangle \cdot \langle k_2, x_2\rangle \\
\widehat{f}(k_1, k_2) &= \int_{G_1}\int_{G_2} \langle -k_1, x_1\rangle\langle -k_2, x_2\rangle f(x_1, x_2)dx_1 dx_2 \\
f(x_1, x_2) &= \frac{1}{C_1 C_2}\int_{\widehat{G_1}}\int_{\widehat{G_2}} \langle k_1, x_1\rangle\langle k_2, x_2\rangle f(k_1, k_2)dk_1 dk_2.
\end{aligned}$$

We end this section with a brief discussion of shifts and the convolution theorem. For a finite group G, we let $\mathbb{C}G$ denote the *group algebra* (or group ring). This is the complex vector space of dimension $|G|$, where each element in G is a basis vector, and with a product given by convolution. The convolution is most easily computed if we write the basis vector for $\lambda \in G$ in the multiplicative form e^λ. Then $f \in \mathbb{C}G$ is represented by the vector $f = \sum_{\lambda\in G} f(\lambda)e^\lambda$ and we obtain the convolution $f*g$ of $f, g \in \mathbb{C}G$ by computing their product and collecting equal terms

$$fg = \sum_{\lambda\in G} f(\lambda)e^\lambda \sum_{\lambda'\in G} g(\lambda')e^{\lambda'} = \sum_{\lambda\in G}(f*g)(\lambda)e^\lambda,$$

where

$$(f*g)(\lambda) = \sum_{\lambda'\in G} f(\lambda')g(\lambda - \lambda').$$

For a continuous group G, we can similarly understand $\mathbb{C}G$ as a function space of complex valued functions on G. Care must be taken in order to define a suitable function space for which convolutions make sense, but these issues will not be discussed in this paper, see [31]. *All results tacitly assume that we define an appropriate function space, e.g. $L^2(G)$, where operations such as convolutions and Fourier transforms are well-defined.*

The continuous convolution is given for $f, g \in \mathbb{C}G$ as

$$(f * g)(x) = \int_{x' \in G} f(x')g(x - x')dx'.$$

Arguably the most important property of the Fourier transform is that it diagonalizes the convolution, i.e.

$$\widehat{(f * g)}(\xi) = \widehat{f}(\xi)\widehat{g}(\xi) \quad \text{for all } \xi \in \widehat{G}.$$

The proof is a straightforward computation. It is worthwhile to notice that it relies upon the property that the group characters are homomorphisms, i.e. $\langle \xi, x+x' \rangle = \langle \xi, x \rangle \langle \xi, x' \rangle$, and upon the translation invariance of the integral.

Similar results hold for more general non-commutative groups, in which case the characters are replaced by group representations. Irreducible group representations are matrix valued group homomorphisms which form an orthogonal basis for $L^2(G)$, stated by the Frobenius theorem for finite G and Peter–Weyl theorem for compact G. Applications of such generalized Fourier transforms in numerical linear algebra are discussed in [2].

9.2.3 Subgroups, lattices and sampling

A subgroup of G, written $H < G$, is a topologically closed subset $H \subset G$ that is closed under the group operations $+$ and $-$.

Definition 9.6 For a subgroup $H < G$ we define the *annihilator subgroup* $H^\perp < \widehat{G}$ as

$$H^\perp = \left\{ k \in \widehat{G} \mid \langle k, h \rangle_G = 1 \text{ for all } h \in H \right\}.$$

Note that if $k, k' \in \widehat{G}$ are such that $k - k' \in H^\perp$, then $\langle k, h \rangle_G = \langle k', h \rangle_G$ for all $h \in H$. This phenomenon is called *aliasing* in signal processing, the two characters corresponding to k and k' are indistinguishable when restricted to H.

Since G is abelian, any subgroup is normal and we can always form the quotient group $K = G/H$, where the elements of K are the cosets $H + g$ with the group operation $(H + g) + (H + g') = H + g + g'$, and the identity element of K is H. Similarly, we can form the quotient $\widehat{H}/H^\perp$, where each coset $H^\perp + k$ consists of a set of characters aliasing on H.

Definition 9.7 A *lattice* in G is a discrete subgroup $H < G$ such that G/H is compact.

An example is $G = \mathbb{R}$, $H = \mathbb{Z}$ and $K = \mathbb{R}/\mathbb{Z} = T$. Also $\mathbb{Z} \times \mathbb{Z} < \mathbb{R} \times \mathbb{R}$ is a lattice. However, $\mathbb{Z} \times 0 < \mathbb{R} \times \mathbb{R}$ is *not* a lattice since the quotient $T \times \mathbb{R}$ is not compact. If G is a finite group, then any $H < G$ is a lattice.

As another example, consider $G = \mathbb{Z}$ with the lattice $2\mathbb{Z} < G$ consisting of all even integers. Theorem 9.2 states that $2\mathbb{Z}$ is isomorphic to $\mathbb{Z}$ as an abstract group, thus it seems natural to define $H = \mathbb{Z}$ and identify H with a subgroup of G via a group homomorphism $\phi_0 \in \hom(H, G)$ given as $\phi_0(j) = 2j$. Similarly the quotient is $K = \mathbb{Z}_2$, identified with the cosets of H in G via $\phi_1 \in \hom(G, K)$ given as $\phi_1(j) = j \bmod 2$, where two elements $j, j' \in G$ belong to the same coset if and only if $\phi_1(j - j') = 0$.

It is in general easy to define which properties of $\phi_1 \in \hom(H, G)$ and $\phi_2 \in (G, K)$ are necessary and sufficient for H to be (isomorphic to) a subgroup of G with quotient (isomorphic to) K. Recall that the kernel and image of $\phi \in \hom(G_1, G_2)$ are defined as $\ker(\phi) = \{ x \in G_1 \mid \phi(x) = 0 \}$ and $\operatorname{im}(\phi) = \phi(G_1) \subset G_2$. A *(co)chain complex* is a sequence $\{G_j, \phi_j\}$ of homomorphisms between abelian groups $\phi_j \in \hom(G_j, G_{j+1})$ such that $\operatorname{im}(\phi_{j+1}) \subset \ker(\phi_j)$ for all j, in other words $\phi_{j+1} \circ \phi_j = 0$ for all j. The sequence is *exact* if $\ker(\phi_{j+1}) = \operatorname{im}(\phi_j)$. A *short exact sequence* is an exact sequence of five terms of the form

$$\mathbf{0} \longrightarrow H \xrightarrow{\phi_0} G \xrightarrow{\phi_1} K \longrightarrow \mathbf{0}\,. \tag{9.1}$$

The leftmost and rightmost arrows are the trivial maps $0 \mapsto 0$ and $K \mapsto 0$. A short exact sequence defines a subgroup $H < G$ with quotient $K = G/H$, or more precisely ϕ_0 is a monomorphism (injective homomorphism) identifying H with a subgroup $\phi_0(H) < G$ and ϕ_1 an epimorphism (surjective homomorphism) identifying $G/\phi_0(H)$ with K. Henceforth we will always define H as a subgroup of G with quotient K by explicitly defining a short exact sequence and the maps ϕ_0 and ϕ_1.

Although this homological algebra point of view is ubiquitous in many areas of pure mathematics, it is not a commonly used language in applied and computational mathematics. However, this presentation is also very important from a computational point of view. First of all, this language allows for a general and unified discussion of sampling, interpolation and Fast Fourier Transforms. Furthermore, from an object oriented programming point of view, it is an advantage to characterize mathematical objects in terms of categorical diagrams. Classes in an object oriented program consist of an internal representation of a certain abstraction as well as an interface defining the interaction and relationship between different objects. Category theory and homological algebra is thus a language which is important in defining classes in object oriented programming. We will not discuss implementations further here, but refer to [1, 14] for examples of this line of ideas within numerical analysis. To understand sampling theory and the FFT in a group language, we need to define *adjoint homomorphisms*, similarly to adjoints of linear operators.

Definition 9.8 Given two LCAs H and G with dual pairings $\langle\cdot,\cdot\rangle_H$ and $\langle\cdot,\cdot\rangle_G$. The adjoint of a homomorphism $\phi \in \hom(H,G)$ is $\widehat{\phi} \in \hom(\widehat{G},\widehat{H})$ defined such that

$$\langle \xi, \phi(x)\rangle_G = \langle \widehat{\phi}(\xi), x\rangle_H \quad \text{for all } \xi \in \widehat{G} \text{ and } x \in H.$$

The following fundamental theorem is proven by standard techniques in homological algebra.

Theorem 9.9 *A short sequence of LCAs*

$$\mathbf{0} \longrightarrow H \xrightarrow{\phi_0} G \xrightarrow{\phi_1} K \longrightarrow \mathbf{0}\,.$$

is exact if and only if the adjoint sequence

$$\mathbf{0} \longleftarrow \widehat{H} \xleftarrow{\widehat{\phi}_0} \widehat{G} \xleftarrow{\widehat{\phi}_1} \widehat{K} \longleftarrow \mathbf{0}\,.$$

is exact.

Note that $\widehat{K}$ is a subgroup of $\widehat{G}$ with quotient $\widehat{G}/\widehat{K} = \widehat{H}$. Furthermore, for any $x \in H$ and for any $k \in \widehat{K}$, we have that $\langle \widehat{\phi}_1(k), \phi_0(h)\rangle_G = \langle k, \phi_1 \circ \phi_0(x)\rangle_K = \langle k, 0\rangle_K = 1$, since the composition of any two adjacent arrows is 0. The exactness of the adjoint sequence implies that if $\xi \in \widehat{G}$ is such that $\langle \xi, \phi_0(h)\rangle_G = 0$ for all $h \in H$, then $\xi = \widehat{\phi}_1(k)$ for some $k \in \widehat{K}$. Thus Theorem 9.9 implies the fundamental dualities

$$\begin{aligned} H &< G, & K &= G/H,\\ \widehat{H} &= \widehat{G}/\widehat{K}, & \widehat{K} &= H^{\perp} < \widehat{G}. \end{aligned}$$

For many computational problems it is necessary to choose representative elements from each of the cosets in the quotient groups G/H and $\widehat{G}/\widehat{K}$. E.g. in sampling theory all characters in a coset $\widehat{K} + \xi \subset \widehat{G}$ alias on H, but physical relevance is usually given to the character $\xi' \in \widehat{K} + \xi$ which is closest to 0 (the lowest frequency mode). Similarly, we often represent G/H by picking a representative from each coset, e.g. $\mathbb{R}/2\pi\mathbb{Z}$ can be represented by $[0, 2\pi) \subset \mathbb{R}$. The quotient map $\phi_1 \colon G \to K$ assigns each coset to a unique element in K, and we need to decide on a right inverse of this map.

Definition 9.10 (Transversal of coset map) Given the short exact sequence (9.1), a function $\sigma: K \to G$ is called a *transversal* of the quotient map $\phi_1: G \to K$ if $\phi_1 \circ \sigma = \mathrm{Id}_K$.

Note that in general we cannot choose σ as a group homomorphism (only if $G = H \times K$), but it can be chosen as a continuous function. In most applications G has a natural norm (e.g. Euclidean distance) and we can choose σ such that the coset representatives are as close to the origin as possible, i.e. such that $||\sigma(k)|| \leq ||\sigma(k) - h||$ for all $h \in \phi_0(H)$. This choice is called a *Voronoi transversal.* It is usually not uniquely defined on the boundary, and the treatment of points on the boundary must be done with some care in many applications. If H is a lattice in a continuous group G, then the closure of the image $\sigma(K) \subset G$ is a polyhedron limited by hyperplanes halfway between 0 and its neighbouring lattice points. In sampling theory one usually picks out the coset representatives for aliasing characters in $\widehat{G}/\widehat{K}$ by letting $\widehat{\sigma}: \widehat{H} \to \widehat{G}$ be the Voronoi transversal of $\widehat{\phi}_0$ with respect to the L^2 norm on $\widehat{G}$.

In the rest of this section, we assume that H, G, K form a short exact sequence as in (9.1), where H is a lattice, i.e. H is discrete and K is compact. Then $\widehat{H}$ is compact and $\widehat{K} = H^{\perp}$ is discrete, so $H^{\perp}$ is a lattice in $\widehat{G}$, called the *reciprocal lattice.* For a function $f \in \mathbb{C}G$, we let $f_H = f \circ \phi_0 \in \mathbb{C}H$ denote the function f *downsampled* to the lattice H. Similarly, $\widehat{f}_{H^{\perp}} = \widehat{f} \circ \widehat{\phi}_1 \in \mathbb{C}H^{\perp}$.

Lemma 9.11 (Poisson summation formula) *Given a lattice $H < G$ with reciprocal lattice $H^{\perp} < \widehat{G}$, there exists a constant C such that*

$$\sum_H f_H = \frac{1}{C} \sum_{H^{\perp}} \widehat{f}_{H^{\perp}}.$$

If G is compact, then $C = \mathrm{vol}(G/H)$. With our normalization of the Fourier transform on $\mathbb{R}$, $C = \mathrm{vol}(G/H)$ also when $G = \mathbb{R}^n$.

Proof Consider the group G/H via its set of coset representatives $V := \sigma(K) \subset G$. The characters of this group are $\{\langle \widehat{\phi}_1(k), \cdot \rangle_G\}_{k \in \widehat{K}}$. Thus by Fourier inversion in G/H there exists a constant C such that

$$f(0) = C \sum_{k \in \widehat{K}} \int_V f(x) \overline{\langle \widehat{\phi}_1(k), x \rangle_G} dx.$$

We write $x \in G$ as $x = y + \phi_0(h)$, where $y \in V$, use $\langle \widehat{\phi}_1(k), \phi_0(h)\rangle \equiv 1$ and the result above to obtain:

$$\begin{aligned}
\sum_{k\in\widehat{K}} \widehat{f}\circ\widehat{\phi}_1(k) &= \sum_{k\in\widehat{K}} \int_G f(x)\overline{\langle\widehat{\phi}_1(k), x\rangle_G} dx \\
&= \sum_{k\in\widehat{K}} \sum_{h\in H} \int_V f(y+\phi_0(h))\overline{\langle\widehat{\phi}_1(k), \phi_0(h)+y\rangle_G} dy \\
&= \sum_{h\in H} \sum_{k\in\widehat{K}} \int_V f(y+\phi_0(h))\overline{\langle\widehat{\phi}_1(k), y\rangle_G} dy \\
&= C\sum_{h\in H} f\circ\phi_0(h).
\end{aligned}$$

If G is compact, we set $f = 1$ and compute $\widehat{f} = \mathrm{vol}(G)\delta_0$. Since $|H| = \mathrm{vol}(G)/\mathrm{vol}(V)$ we find $C = \mathrm{vol}(V)$. The constant is computed for $G = \mathbb{R}^n$ by considering the Fourier transform of $f(x) = e^{-x^T x}$, which under appropriate scaling, is invariant under the Fourier transform on $\mathbb{R}^n$. □

9.2.4 Heisenberg groups and the FFT

More material on topics related to this section is found in [4, 5, 35].

We can act upon $f \in \mathbb{C}G$ with a time-shift $S_x f(t) := f(t+x)$ and with a frequency shift $\chi_\xi f(t) := \langle\xi, t\rangle f(t)$. These two operations are dual under the Fourier transform, but do not commute:

$$\widehat{S_x f}(\xi) = \chi_x \widehat{f}(\eta) \tag{9.2}$$

$$\widehat{\chi_\xi f}(\eta) = S_{-\xi}\widehat{f}(\eta) \tag{9.3}$$

$$\left(S_x \chi_\xi f\right)(t) = \langle\xi, x\rangle \cdot \left(\chi_\xi S_x f\right)(t). \tag{9.4}$$

The full (non-commutative) group generated by time and frequency shifts on $\mathbb{C}G$ is called the *Heisenberg group* of G.

The Heisenberg group of $\mathbb{R}^n$ is commonly defined as the multiplicative group of matrices of the form

$$\begin{pmatrix} 1 & x^T & s \\ 0 & I_n & \xi \\ 0 & 0 & 1 \end{pmatrix},$$

where $\xi, x \in \mathbb{R}^n$, $s \in \mathbb{R}$. This group is isomorphic to the semidirect product $\mathbb{R}^n \times \mathbb{R}^n \rtimes \mathbb{R}$ where

$$(\xi', x', s') \cdot (\xi, x, s) = (\xi' + \xi, x' + x, s' + s + x'^T \xi).$$

We prefer to instead consider $\mathbb{R}^n \times \mathbb{R}^n \rtimes \mathbb{T}$ (where $\mathbb{T}$ is the multiplicative group consisting of $z \in \mathbb{C}$ such that $|z| = 1$) with product

$$(\xi', x', z') \cdot (\xi, x, z) = (\xi' + \xi, x' + x, z'ze^{2\pi i x'^T \xi}).$$

More generally:

Definition 9.12 For an LCA G we define the Heisenberg group $\mathcal{H}_G = \widehat{G} \times G \rtimes \mathbb{T}$ with the semidirect product

$$(\xi', x', z') \cdot (\xi, x, z) = (\xi' + \xi, x' + x, z' \cdot z \cdot \langle \xi, x' \rangle).$$

We define a *left action* $\mathcal{H}_G \times \mathbb{C}G \to \mathbb{C}G$ as follows

$$(\xi, x, z) \cdot f = z \cdot \chi_\xi S_x f. \tag{9.5}$$

To see that this defines a left action, we check that $(0, 0, 1) \cdot f = f$ and

$$(\xi', x', z') \cdot ((\xi, x, z) \cdot f) = ((\xi', x', z') \cdot (\xi, x, z)) \cdot f.$$

Lemma 9.13 *Let $\mathcal{H}_G = \widehat{G} \times G \rtimes \mathbb{T}$ and $\mathcal{H}_{\widehat{G}} = G \times \widehat{G} \rtimes \mathbb{T}$ act upon $f \in \mathbb{C}G$ and $\widehat{f} \in \mathbb{C}\widehat{G}$ as in (9.5). Then*

$$\mathcal{F}((\xi, x, z) \cdot f) = z \cdot \langle -\xi, x \rangle \cdot \chi_x S_{-\xi} \widehat{f} = (x, -\xi, z \cdot \langle -\xi, x \rangle) \cdot \widehat{f}$$

Proof This follows from (9.2)–(9.4). □

We will henceforth assume that H, G, K forms a short exact sequence as in (9.1), with H discrete and K compact.

Definition 9.14 (Weil–Brezin map) The *Weil–Brezin map* $\mathcal{W}_G^H$ is defined for $f \in \mathbb{C}G$ and $(\xi, x, s) \in \mathcal{H}_G$ as

$$\mathcal{W}_G^H f(\xi, x, z) = \sum_H ((\xi, x, z) \cdot f)_H \cdot$$

A direct computation shows that the Weil–Brezin map satisfies the following symmetries for all $(h', h, 1) \in H^\perp \times H \times 1 \subset \mathcal{H}_G$ and all $z \in \mathbb{T}$:

$$\mathcal{W}_G^H f((h', h, 1) \cdot (\xi, x, s)) = \mathcal{W}_G^H f(\xi, x, s) \tag{9.6}$$

$$\mathcal{W}_G^H f(\xi, x, z) = z \cdot \mathcal{W}_G^H f(\xi, x, 1). \tag{9.7}$$

Lemma 9.15 *$\Gamma = H^\perp \times H \times 1$ is a subgroup of $\mathcal{H}_G$. It is not a normal subgroup, so we cannot form the quotient group. However, as a manifold the set of right cosets is*

$$\Gamma \backslash \mathcal{H}_G = \widehat{H} \times K \times \mathbb{T}.$$

The Heisenberg group has a right and left invariant volume measure given by the direct product of the invariant measures of $\widehat{G}$, G and $\mathbb{T}$. Thus we can define the Hilbert spaces $L^2(\mathcal{H}_G^H)$ and $L^2(\widehat{H} \times K \times \mathbb{T})$. By Fourier decomposition in the last variable (z-transform), $L^2(\widehat{H} \times K \times \mathbb{T})$ splits into an orthogonal sum of subspaces $\mathcal{V}_k$ for $k \in \mathbb{Z}$, consisting of those $g \in L^2(\widehat{H} \times K \times \mathbb{T})$ such that

$$g(\xi, x, z) = z^k g(\xi, x, 1) \quad \text{for all } z = e^{2\pi i \theta}.$$

It can be verified that $\mathcal{W}_G^H$ is unitary with respect to the L^2 inner product. Together with (9.6)–(9.7) this implies:

Lemma 9.16 *The Weil–Brezin map is a unitary transform*

$$\mathcal{W}_G^H \colon L^2(G) \to \mathcal{V}_1 \subset L^2(\widehat{H} \times K \times \mathbb{T}).$$

Note that the Weil–Bezin map on $\widehat{G}$, with respect to the reciprocal lattice $H^\perp$, is a unitary map

$$\mathcal{W}_{\widehat{G}}^{H^\perp} : L^2(G) \to \mathcal{V}_1 \subset L^2(K \times \widehat{H} \times \mathbb{T}).$$

The Poisson summation formula (Lemma 9.11) together with Lemma 9.13 implies that these two maps are related via

$$\mathcal{W}_G^H f(\xi, x, z) = \mathcal{W}_{\widehat{G}}^{H^\perp} \widehat{f}(x, -\xi, z \cdot \langle \xi, x \rangle).$$

Defining the unitary map $J \colon L^2 \subset L^2(\widehat{H} \times K \rtimes \mathbb{T}) \to L^2(K \times \widehat{H} \times \mathbb{T})$ as

$$Jf(x, -\xi, z \cdot \langle \xi, x \rangle) = f(\xi, x, z),$$

we obtain the following fundamental theorem:

Theorem 9.17 (Weil–Brezin factorization) *Given an LCA G and a lattice $H < G$. The Fourier transform on G factorizes in a product of three unitary maps*

$$\mathcal{F}_G = \left(\mathcal{W}_{\widehat{G}}^{H^\perp}\right)^{-1} \circ J \circ \mathcal{W}_G^H. \tag{9.8}$$

The Zak transform. Given a lattice $H < G$ and transversals $\sigma \colon K \to G$ and $\widehat{\sigma} \colon \widehat{H} \to \widehat{G}$. The *Zak transform* is defined as

$$\mathcal{Z}_G^H f(\xi, x) := \mathcal{W}_G^H f(\xi, x, 1) \quad \text{for } \xi \in \widehat{\sigma}(\widehat{H}),\ x \in \sigma(K).$$

The Zak transform can be computed as a collection of Fourier transforms on H of f shifted by x, for all $x \in \sigma(K)$. The definition of the Fourier transform yields:

$$\mathcal{Z}_G^H f(-\xi, x) = \mathcal{F}_H\left((S_x f)_H\right)(\widehat{\phi}_0(\xi)).$$

We see that the Zak transform is invertible when $\mathcal{Z}_G^H f(-\xi, x)$ is computed for all $\xi \in \widehat{\sigma}(\widehat{H})$ and all $x \in \sigma(K)$. Written in terms of the Zak transform, the Weil–Brezin factorization (9.8) becomes

$$\mathcal{Z}_{\widehat{G}}^{H^{\perp}} \widehat{f}(x, \xi) = \langle \xi, x \rangle \mathcal{Z}_G^H f(-\xi, x). \tag{9.9}$$

The factor $\langle \xi, x \rangle$ is called a *twiddle factor* in the computational FFT literature.

In the special case where $G = H \times K$, then also $K < G$ and $\widehat{H} = K^{\perp}$. Thus in this case it is possible to choose σ and $\widehat{\sigma}$ as group homomorphisms, resulting in $\langle \xi, x \rangle \equiv 1$ for all $\xi \in \widehat{\sigma}(\widehat{H})$ and $x \in \sigma(K)$. This choice is called a *twiddle free* factorization. However, by other choices of the transversals, the twiddle factors also enter into the formula in this case.

Due to the symmetries (9.6)–(9.7), the Weil–Brezin map is trivially recovered from the Zak transform. The Zak transform is the practical way of computing the Weil–Brezin map and its inverse. However, since the invertible Zak transform cannot be defined canonically, independently of the transversals σ and $\widehat{\sigma}$, the Weil–Brezin formulation is more fundamental.

The Fast Fourier Transform. Cooley–Tukey style FFT algorithms are based on recursive use of (9.9), where a Fourier transform on G is computed by a collection of Fourier transforms on H composed with inverse Fourier transforms on $H^{\perp}$. We choose transversals σ and $\widehat{\sigma}$. If G is finite, then $\widehat{\sigma}(\widehat{H})$ and $\sigma(K)$ are finite. The Cooley–Tukey factorization follows from (9.9):

- For each $x \in \sigma(K)$ compute:
$$\mathcal{Z}_G^H f(-\xi, x) = \mathcal{F}_H(S_x f)(\widehat{\phi}_0(\xi)) \quad \text{for all } \xi \in \widehat{\sigma}(\widehat{H})\ .$$
- For each $\xi \in \widehat{\sigma}(\widehat{H})$ compute:
$$\widehat{f}(\xi + \widehat{\phi}_1(\kappa)) = \mathcal{F}_{H^{\perp}}^{-1}\left(\langle \xi, \sigma(\cdot) \rangle \mathcal{Z}_G^H f(-\xi, \sigma(\cdot))\right)(\kappa) \quad \text{for all } \kappa \in H^{\perp},$$
where the inverse Fourier transform $\mathcal{F}_{H^{\perp}}^{-1}$ is with respect to the variable $\cdot \in K$.

The Fast Fourier Transform is obtained by recursive application of this splitting. This general formulation allows for Cooley–Tukey kind FFTs based on any decomposition of G with respect to a lattice H. In particular this is useful for functions with symmetries, in which case it is important to choose lattices H that preserve the symmetries in order to take advantage of all the symmetries in the FFT. We return to this issue in Section 9.3.4.

Shannon's sampling theorem. By setting $x = 0$ in (9.9), we obtain the important dual relationship between downsampling and periodization:

$$\mathcal{F}_H(f_H)(\widehat{\phi}_0(\xi)) = \sum_{k \in \widehat{\phi}_1(\widehat{K})} \mathcal{F}_G(f)(k+\xi). \tag{9.10}$$

A function $f \in \mathbb{C}G$ is *band limited* with respect to the reciprocal lattice $H^\perp$ if its Fourier transform is zero outside the Voronoi polyhedron, i.e. if $\mathrm{supp}(\widehat{f}) \subset \widehat{\sigma}(\widehat{H})$, where $\widehat{\sigma}$ is a Voronoi transversal of $\widehat{\phi}_0$. If f is band limited, the terms on the right hand side of (9.10) are zero for $k \neq 0$ and $\xi \in \widehat{\sigma}(\widehat{H})$. This yields

$$\mathcal{F}_G(f)(\xi) = \mathcal{F}_H(f_H)(\widehat{\phi}_0(\xi)).$$

Thus we obtain Shannon's celebrated result that a band limited f can be exactly recovered from its downsampling f_H.

Lattice rules. For general functions $f \in \mathbb{C}G$, the error between the Fourier transform of the true and the sampled function is given as

$$\mathcal{F}_H(f_H)(\widehat{\phi}_0(\xi)) - \mathcal{F}_G(f)(\xi) = \sum_{k \in \widehat{\phi}_1(\widehat{K})\backslash\{0\}} \mathcal{F}_G(f)(k+\xi).$$

The game of Lattice rules is, given f with specific properties, to find a lattice $H < G$ such that the error is minimised. We now assume that the original domain is periodic $G = T^n$. Lattice rules are designed such that the nonzero points in $H^\perp$ neighbouring 0 are pushed as far out as possible with respect to a given norm, depending on f. If f is spherically symmetric, H should be chosen as a *densest lattice packing* (with respect to the 2-norm) [10], e.g. hexagonal lattice in $\mathbb{R}^2$ and face centred cubic packing in $\mathbb{R}^3$ (as the orange farmers know well). In dimensions up to 8, these are given by certain root lattices [29]. The savings, compared to standard tensor product lattices, are given by the factors 1.15, 1.4, 2.0, 2.8 4.6, 8.0 and 16.0 in dimensions $n = 2, 3, \ldots, 8$. This is important, but not dramatic, e.g. a camera with 8.7 megapixels arranged in a hexagonal lattice has approximately the same sampling error as a 10 megapixel camera with a standard square pixel distribution. However, these alternative lattices have other attractive features, such as larger spatial symmetry groups, yielding more isotropic discretizations.

A hexagonal lattice picture can be rotated more uniformly than a square lattice picture.

Dramatic savings can be obtained for functions belonging to the *Korobov spaces.* This is a common assumption in much work on high dimensional approximation theory. Korobov functions are functions whose Fourier transforms have their energy concentrated along the axis directions in $\widehat{G}$, the so-called hyperbolic cross mass distribution. Whereas the tensor product lattice with $2d$ points in each direction contains $(2d)^n$ lattice points in T^n, the optimal lattice with respect to the Korobov norm contains only $\mathcal{O}(2^n d(\log(d))^{n-1})$ points, removing exponential dependence on d.

The group theoretical understanding of lattice rules makes software implementation very clean and straightforward. In [30], numerical experiments are reported on lattice rules for FFT-based spectral methods for PDEs. Note that whereas the choice of transversal $\widehat{\sigma}\colon \widehat{H} \to \widehat{G}$ is irrelevant for lattice integration rules, it is essential for pseudo-spectral derivation. The Laplacian $\nabla^2 f$ is computed on $\widehat{G}$ as $\widehat{f}(\xi) \mapsto c|\xi|^2\widehat{f}(\xi)$, whereas the corresponding computation on $\widehat{H}$ must be done as $\mathcal{F}_H(f_H)(\eta) \mapsto c|\widehat{\sigma}(\eta)|^2\mathcal{F}_H(f_H)(\eta)$ for $\eta \in \widehat{H}$, and we must choose the Voronoi transversal to minimise aliasing errors.

Polyhedral Dirichlet kernels. The theoretical understanding of lattice sampling rules depends on the analytical properties of *polyhedral Dirichlet kernels.* Let $\widehat{\sigma}\colon \widehat{H} \to \widehat{G}$ be the Voronoi transversal. The perfect low-pass filter $\widehat{\mathcal{D}}_H \in \mathbb{C}\widehat{G}$ is defined as

$$\widehat{\mathcal{D}}_H(\xi) = \begin{cases} 1 & \text{if } \xi \in \widehat{\sigma}(\widehat{H}) \\ 0 & \text{otherwise} \end{cases} .$$

The polyhedral Dirichlet kernel $\mathcal{D}_H \in \mathbb{C}G$ is defined as

$$\mathcal{D}_H = \mathcal{F}_G^{-1}(\widehat{\mathcal{D}}_H).$$

This function plays the same role as the classical Dirichlet kernel in the 1-dimensional sampling theory, e.g. low-pass reconstruction of a down sampled function is done by convolution with $\mathcal{D}_H$. Detailed analysis of these functions is done in [34, 36]. In particular, it is important that they in general have Lebesque constants scaling like $\mathcal{O}(\log^n(N))$, where n is the dimension of G, and N measures the number of sampling points in H.

9.2.5 Fourier analysis on non-commutative groups

In this section we will briefly discuss to what degree Fourier techniques on non-commutative groups have relevance in computations. We will be much less detailed than in the previous section, since this material is covered in detail elsewhere, e.g. [3, 2, 29].

A starting point of the LCA discussion was the definition of the group ring $\mathbb{C}G$ and the existence of a translation invariant measure, which led to convolutions in the group ring. For non-commutative groups the situation is a bit more complicated, since invariance with respect to left and right translations might not yield the same measure. Groups for which there exist a (unique up to scaling) measure which is both left and right invariant is called *unimodular.* For such groups a lot of the previous theory carries over, with some modifications. Groups which are not unimodular are considerably more complicated and will not be discussed here.

Important examples of unimodular groups are:

- Abelian groups.
- Finite groups.
- Compact groups.
- Semidirect product of compact and abelian groups, e.g. the Euclidean motion group consisting of translations and rotations.
- Semisimple and nilpotent Lie groups.

Finite groups. Let $\mathbb{C}G$ denote the group ring, the complex vector space of dimension $|G|$, where each element in G is a basis vector, so, as before, $f \in \mathbb{C}G$ is given as $f = \sum_{g\in G} f(g)g$, where $f(g) \in \mathbb{C}$. The right and left invariant Haar measure is given as the sum over the elements

$$\int_G f d\mu = \sum_{x\in G} f(x).$$

The product in G yields a convolution product in $\mathbb{C}G$

$$(f * g)(y) = \sum_{x\in G} f(x)g(x^{-1}y) = \sum_{x\in G} f(yx)g(x^{-1}).$$

This is, however, not a commutative product on $\mathbb{C}G$, $f * g \neq g * f$. In the abelian case, the Fourier transform diagonalizes the convolution because the exponential basis consists of group homomorphisms (into $\mathbb{T}$). In the non-commutative case, it cannot be possible to diagonalize the convolution using just $\hom(G, \mathbb{T})$ because the convolution is not commutative. The idea of Schur and Frobenius in the late 19th century was to look for a basis for $\mathbb{C}G$ in terms of *group representations*, defined as elements of $\hom(G, U(n))$, where $U(n)$ is the set of unitary $n \times n$ matrices. (Note that $U(1) = \mathbb{T}$.) An n-dimensional representation is thus a function $R \colon G \to U(n)$ satisfying $R(xy) = R(x)R(y)$ and $R(x^{-1}) = R(x)^{-1} = R(x)^{\dagger}$, where $R(x)^{\dagger}$ is the complex conjugate transpose. Let $d_R = n$ denote the dimension of the representation. For each representation R, we may define Fourier coefficients of a function $f \in \mathbb{C}G$ as a complex $d_R \times d_R$ matrix defined as

$$\widehat{f}(R) = \sum_{x \in G} f(x) R(x)^{\dagger}.$$

A computation using the homomorphism property and shift invariance of the sum, shows that the representations may be used to (block-) diagonalize the convolution:

$$\widehat{f * g}(R) = \widehat{g}(R)\widehat{f}(R).$$

However, we need a basis for $\mathbb{C}G$, and we need an inversion formula for this generalised Fourier transform.

The concepts of equivalent and reducible representations are crucial for constructing a suitable basis. Two representations R and $\tilde{R}$ are equivalent if there exists an invertible matrix V such that $R(x) = V\tilde{R}(x)V^{-1}$ for all x. A representation is reducible if it is equivalent to a representation which is block diagonal. In that case the representation can be seen as a direct sum of smaller representations (one for each diagonal block). Frobenius found that there always exists a complete list of non-equivalent irreducible representations which forms an orthogonal basis for $\mathbb{C}G$.

Theorem 9.18 (Frobenius) *For a finite group G there exists a complete list of non-equivalent irreducible representations $\mathcal{R} = \{R_1, \ldots, R_k\}$ such that $\sum_{R \in \mathcal{R}} d_R^2 = |G|$. Define the generalized Fourier transform of $f \in \mathbb{C}G$ as*

$$\widehat{f}(R) = \sum_{x \in G} f(x) R(x)^{\dagger} \quad \text{for all } R \in \mathcal{R}.$$

Then f is reconstructed by the formula

$$f(x) = \frac{1}{|G|} \sum_{R \in \mathcal{R}} d_R \operatorname{trace}(\widehat{f}(R) R(x)).$$

As an example, we consider the computation of the convolution of

$f, g \in \mathbb{C}G$ when $G < O(3)$ is the icosahedral group, the collection of the 120 orthogonal matrices which leave the icosahedron in $\mathbb{R}^3$ invariant. This group has a complete list of irreducible representations of dimensions $\{1, 1, 3, 3, 3, 3, 4, 4, 5, 5\}$. A direct computation of the convolution involves 120^2 multiplications. Instead, computing the convolution in Fourier space involves the multiplication of matrices of size $1, 1, 3, 3, \ldots$, which requires only 120 multiplications, saving a factor of 120. For the computation of matrix exponentials and eigenvalues, the savings are more dramatic; a direct computation costs 120^3 operations, while the equivalent computation in Fourier space costs $2 + 4 * 3^3 + 2 * 4^3 + 2 * 5^3$ operations, which is cheaper by a factor of about 3500.

A source of computational problems leading to group convolutions is linear problems with spatial symmetries. Given a linear operator $\mathcal{L}$ which commutes with a finite group of isometries acting upon the domain (e.g. the Laplacian on the sphere commutes with any group of isometries, e.g. the discrete icosahedral group). Let $\mathcal{L}$ be discretized in a symmetry preserving manner, such that the discrete L commutes with the isometries in G. Then L can be described as a block convolution, i.e. L belongs to a group ring of the form $\mathbb{C}^{m \times m}G$. The blocks represent the interaction between different orbits of the action of G on the domain. As an example, if the space of spherical functions is discretized with 12,000 degrees of freedom, the full space splits into about $12,000/120 = 100$ different orbits under the action of the icosahedral group. The Laplacian can then be represented as a block convolution in $\mathbb{C}^{100 \times 100}G$. Under the generalized Fourier transform, the matrix becomes block diagonal, with blocks of sizes $100 d_R$ for $d_R \in \{1, 1, 3, 3, \ldots, 5\}$.

The use of the generalized Fourier transform is important in various computational tasks, such as eigenvalue problems, solutions of linear equations and computations of matrix exponentials. Experience shows that symmetry preserving discretizations and algorithms are not only much faster than direct algorithms, but they are also often more accurate, since preservation of symmetry tends to diminish the effect of numerical round-off errors.

Compact groups. Compact groups, such as, for example, the group of rotations SO(3), are in many respects quite similar to finite groups. There exists a bi-invariant Haar measure and a space of functions $L^2(G)$ with a convolution product. The Peter–Weyl theorem guarantees that there exist a discrete, infinite family of irreducible unitary representations forming an orthogonal basis for $L^2(G)$. The Fourier transform becomes an integral over G and the inversion formula is similar to the finite case, although the sum here is over an infinite list $\mathcal{R}$.

The representation theory of orthogonal groups has numerous applications in computational science and technology, an example being recent work on Cryo-Electron microscopy [18, 19]. The basic problem here is the reconstruction of a 3D molecular structure from a large collection of 2D projections of the molecule seen from unknown angles. Representation theory provides an important tool to analyze numerical algorithms for this problem.

Euclidean motions. The Euclidean motion group of translations and rotations on $\mathbb{R}^3$ is important in many technological applications. The group $E(3) = \mathrm{SO}(3) \rtimes \mathbb{R}^3$ is the semidirect product of a compact group and an abelian group. Such groups are always unimodular, and the representation theory is relatively simple. The irreducible representations on $E(3)$ are induced from the representations of the compact part, $SO(3)$, by a standard method called the method of small groups [33]. An example of an application of Fourier analysis on $E(3)$ is the problem of medical image registration: *Find the Euclidean motion which best matches two different 3-D images of an object.* This can be phrased as the question of computing the maximum of the cross correlation (or phase correlation) of the two images. The cross correlation is very similar to a convolution and can be cheaply computed in Fourier space.

9.3 Multivariate Chebyshev polynomials in computations

Univariate (classical) Chebyshev polynomials are ubiquitous in numerical analysis and computational science, due to their in many ways optimal approximation properties and the tight relationship between Chebyshev approximations and fast cosine transforms. First and second kind Chebyshev polynomials $\{T_k\}_{k=0}^{\infty}$ and $\{U_k\}_{k=0}^{\infty}$ are defined as

$$\begin{aligned} x &= \cos(\theta) \\ T_k(x) &= \cos(k\theta) \\ U_k(x) &= \sin((k+1)\theta)/\sin(\theta). \end{aligned}$$

In this section we will discuss the connection between Chebyshev approximations and group theory. Once the group theoretical view is established, it will become clear that there exist certain interesting multivariate generalizations of Chebyshev polynomials. These share most of the favorable properties of the univariate polynomials, and they are orthogonal on domains related to triangles and simplices. Bivariate Chebyshev polynomials were constructed independently by Koornwinder [25] and Lidl [26] by folding exponential functions. Multidimensional generalizations (the A_2 family) appeared first in [13]. In [21] a general folding construction was presented. Characterization of such polynomials as eigenfunctions of differential operators is found in [6, 25].

Our interest in these polynomials originates from their potential applications in computational approximation methods, in particular spectral element methods and multidimensional quadrature. We have developed the theory of their discrete orthogonality, triangular based Clenshaw–Curtis type quadrature formulae, recursion formulae for computing spectral derivations as well as software for the application of multivariate Chebyshev polynomials in approximation, quadrature and PDE solution. Our exposition here will aim at giving an overview of the main ideas. More detailed presentations are found in [29, 32, 9].

Spectral element methods are computational techniques for solving PDEs where the domain of the equation is divided into a fixed collection of regularly shaped subdomains. On each subdomain a high order polynomial space is constructed, and a global solution is obtained by patching together local solutions, either in a strong sense by imposing continuity conditions across subdomain boundaries, or in a weak sense by variational formulations (discontinuous Galerkin methods). The advantage of spectral element methods compared to its competitors (finite elements, finite differences and finite volume methods) is the phenomenon called *spectral convergence*. When an analytic function is approximated by an Nth order polynomial, one may achieve errors decaying as e^{-N}. Thus spectral methods are particularly attractive when high accuracy is important.

The drawbacks of spectral element methods are that high order polynomial approximations must be constructed with care. Jim Wilkinson famously demonstrated that high order polynomial interpolation in equi-spaced points is a highly unstable process, due to the fact that the Lebesgue constant of equi-spaced interpolation points grows exponentially in N. Interpolation in Chebyshev zeros, or Chebyshev extremal points, is on the other hand near optimally stable, as the Lebesque constant in such points grows as $\mathcal{O}(\log(N))$. Another problem with spectral element methods is inflexibility with respect to sub-domain divisions.

High order polynomial approximations are easy to construct on rectangular subdomains (by tensor products of univariate polynomials), but high order approximation theory based on triangular and tetrahedral subdivisions is far less developed. The most common practice is therefore rectangular subdivision methods. Triangular and tetrahedral subdivision schemes are far more flexible, if they can be implemented in an efficient and stable manner. A singular mapping technique from squares to triangles [12] is a possible solution, but has drawbacks in breaking of triangular symmetries as well as other problems. Nodal spectral Galerkin methods are another approach where good collocation nodes are computed by numerical optimization (e.g. Fekete points). But in this approach one has no direct connection to Fourier analysis, and fast transforms are not available [17, 20]. This makes spectral element methods based on multivariate Chebyshev polynomials an attractive alternative. We start with a discussion of particular eigenfunctions of the Laplacian on simplices.

9.3.1 What is the sound of a triangular drum?

Bases for high order approximation spaces are usually obtained as eigenfunctions of Sturm–Liouville problems, truncated to a given order. Can we find Sturm–Liouville problems on triangles and simplices, that yield good approximation spaces? It is known that the eigenfunctions of the Laplacian (with Dirichlet or Neumann boundary conditions) can be explicitly constructed on certain triangular domains in 2D and some particular simplices in all higher dimensions. An illustrative example is the construction of Laplacian eigenfunctions on an equilateral triangle, with Dirichlet or Neumann boundary conditions.

The equilateral triangle has the particular property that if we set up a kaleidoscope with three mirrors at the three edges, then the reflections of the triangle tile the plane in a periodic pattern, shown as the shaded domain in the right part of Figure 9.2 (labelled A_2). Without loss of generality, we assume that the triangle has corners in the origin $[0;0]$, $\lambda_1 = [1/\sqrt{2};1/\sqrt{6}]$ and $\lambda_2 = [0;\sqrt{2}/\sqrt{3}]$. Let $\{s_j\}_{j=1}^3$ denote reflections of $\mathbb{R}^2$ about the edges of the triangle. Let $\widetilde{W}$ denote the full group of isometries of $\mathbb{R}^2$ generated by $\{s_j\}_{j=1}^3$. This is an example of a *crystallographic group*[3], a group of isometries of $\mathbb{R}^d$ where the subgroup

[3] More specifically it is an *affine Weyl group*, to be defined below.

of translations form a lattice in $\mathbb{R}^d$. From this fact we will derive the Laplacian eigenfunctions on the triangle, which we denote by $\triangle$.

The translation lattice of $\widetilde{W}$ is $L = \mathrm{span}_{\mathbb{Z}}\{\alpha_1, \alpha_2\} < \mathbb{R}^2$ generated by the vectors $\alpha_1 = (\sqrt{2}, 0)$ and $\alpha_2 = (-1/\sqrt{2}, \sqrt{3}/\sqrt{2})$. The unit cell of L can be taken as either the rhombus spanned by α_1 and α_2 or the hexagon $\hexagon$ indicated in Figure 9.2. The hexagon is the Voronoi cell of the origin in the lattice L, its interior consists of the points in $\mathbb{R}^2$ that are closer to the origin than to any other lattice points. As a first step in our construction of triangular eigenfunctions, we consider the L-periodic eigenfunctions of the Laplacian ∇^2. Since $(\lambda_j, \alpha_k) = \delta_{j,k}$, the reciprocal lattice is $L^\perp = \mathrm{span}_{\mathbb{Z}}\{\lambda_1, \lambda_2\}$ and the periodic eigenfunctions are

$$\nabla_t^2 \langle \lambda, t\rangle = -(2\pi)^2 \|\lambda\|^2 \langle \lambda, t\rangle \quad \text{for all } \lambda \in L^\perp,\ t \in \mathbb{R}^2/L,$$

where $\langle \lambda, t\rangle = e^{2\pi i(\lambda,t)}$ is the dual pairing on $\mathbb{R}^2$. We continue to find the Laplacian eigenfunctions on $\triangle$ by folding the exponentials. Let $W < \widetilde{W}$ be the subgroup which leaves the origin fixed (the symmetries of $\hexagon$):

$$W = \{e, s_1, s_2, s_1s_2, s_2s_1, s_1s_2s_1\},$$

where e is the identity and s_i, $i \in \{1, 2\}$, act on $v \in \mathbb{R}^2$ as $s_i v = v - 2(\alpha_i, v)/(\alpha_i, \alpha_i)$. We define even and odd (cosine and sine-type) foldings of the exponentials

$$c_\lambda(t) = \frac{1}{|W|} \sum_{w\in W} \langle \lambda, wt\rangle = \frac{1}{|W|} \sum_{w\in W} \langle w^T\lambda, t\rangle$$

$$s_\lambda(t) = \frac{1}{|W|} \sum_{w\in W} \det(w) \langle \lambda, wt\rangle = \frac{1}{|W|} \sum_{w\in W} \det(w) \langle w^T\lambda, t\rangle,$$

where, in our example, $|W| = 6$. The Laplacian commutes with any isometry, in particular $\nabla^2(f \circ w) = (\nabla^2 f) \circ w$ for $w \in W$. Hence, the reflected exponentials and the functions $c_\lambda(t)$ and $s_\lambda(t)$ are eigenfunctions with the same eigenvalue:

Lemma 9.19 *The eigenfunctions of ∇^2 on the equilateral triangle $\triangle$ with Dirichlet ($f = 0$) and Neumann boundary conditions ($\nabla f \cdot \vec{n} = 0$) are given respectively as*

$$\nabla_t^2 c_\lambda(t) = -(2\pi)^2 \|\lambda\|^2 c_\lambda(t)$$
$$\nabla_t^2 s_\lambda(t) = -(2\pi)^2 \|\lambda\|^2 s_\lambda(t),$$

for $\lambda \in \mathrm{span}_{\mathbb{N}}\{\lambda_1, \lambda_2\}$, the set of all non-negative integer combinations of $\lambda_1 = [1/\sqrt{2}; 1/\sqrt{6}]$ and $\lambda_2 = [0; \sqrt{2}/\sqrt{3}]$.

The set $\text{span}_{\mathbb{N}}\{\lambda_1, \lambda_2\} \subset L^{\perp}$ contains exactly one point from each W-orbit, and is called the positive *Weyl chamber.* An important question is how good (or bad!) these eigenfunctions are as bases for approximating analytic functions on $\triangle$. It is well known, from the univariate case, that the similar construction yielding the eigenfunctions $\cos(k\theta)$ and $\sin(k\theta)$ on $[0, \pi]$ do *not* give a basis converging at a super-algebraic speed. For an analytic function $f(\theta)$ where $f(0) = f(\pi) = 0$, the even 2π-periodic extension is piecewise smooth, with only C^0 continuity at $\theta \in \{0, \pi\}$. Hence the Fourier-cosine series converges only as $\mathcal{O}(k^{-2})$. There are several different ways to achieve the desired $\mathcal{O}(\exp(-ck))$ spectral convergence rate for analytic functions. One possible solution is to approximate $f(\theta)$ in a *frame* (not linearly independent) consisting of both $\{\cos(k\theta)\}_{k\in\mathbb{Z}}$ and $\{\sin(k\theta)\}_{k\in\mathbb{Z}^+}$ [22]. Another possibility is to employ a change of variable $x = \cos(\theta)$, yielding (univariate) Chebyshev polynomials with spectral convergence. Here we will discuss a generalization of the latter approach leading to the multivariate Chebyshev polynomials.

9.3.2 Through the kaleidoscope

Recall that the main trick for finding eigenfunctions of the Laplacian on the triangle was in using the fact that the domain is a polyhedron with the property that the group generated by the boundary reflections is a crystallographic group, defined as a group of isometries on $\mathbb{R}^n$, such that the subgroup of translations form a lattice $L < \mathbb{R}^n$. Crystallographic groups is a classical topic of mathematics, physics and chemistry. A classification of such groups starts with the fundamental *crystallographic restriction*: For any crystallographic group, the only allowed rotations are 2-fold, 3-fold, 4-fold and 6- fold[4]. A general group generated just by reflections, as in a kaleidoscope, is called a *Coxeter* group, and the Coxeter groups which comply with the crystallographic restriction are called affine Weyl groups. These are classified in terms of their root systems, where the roots are orthogonal to the mirrors passing through the origin.

Root systems. Let $\mathcal{V}$ be a finite dimensional real Euclidean vector space with standard inner product $(\cdot, \cdot)$. The construction above can be generalized to all those simplices $\triangle \subset \mathcal{V}$ with the property that

[4] The 5-fold symmetry is not allowed in proper crystals, but is seen in quasicrystals, related to Penrose tilings. This discovery was the topic of the 2011 Nobel prize in chemistry.

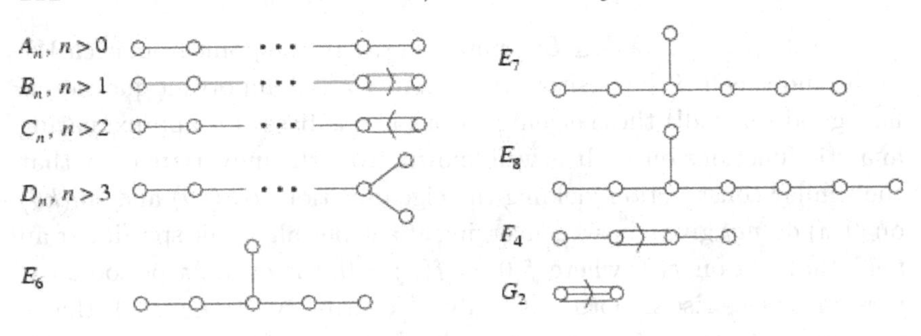

Figure 9.1 Dynkin diagrams for irreducible root systems

the group of isometries $\widetilde{W}$ generated by reflecting $\triangle$ about its faces is a crystallographic group. All such simplices are determined by a *root system*, a set of vectors in $\mathcal{V}$ which are perpendicular to the reflection planes of W passing through the origin. We review some basic definitions and results about root systems. For more details we refer to [8].

Definition 9.20 A **root system** in $\mathcal{V}$ is a finite set Φ of non-zero vectors, called roots, that satisfy:

1. The roots span $\mathcal{V}$.
2. The only scalar multiples of a root $\alpha \in \Phi$ that belong to Φ are α and $-\alpha$.
3. For every root $\alpha \in \Phi$, the set Φ is invariant under reflection through the hyperplane perpendicular to α. I.e. for any two roots α and β, the set Φ contains the reflection of β,
$$s_\alpha(\beta) := \beta - 2\frac{(\alpha, \beta)}{(\alpha, \alpha)}\alpha \in \Phi.$$
4. (Crystallographic restriction): For any $\alpha, \beta \in \Phi$ we have
$$2\frac{(\alpha, \beta)}{(\alpha, \alpha)} \in \mathbb{Z}.$$

Condition 4. implies that the obtuse angle between two different reflection planes must be either 90°, 120°, 135° or 150°. The *rank* of the root system is the dimension d of the space $\mathcal{V}$. Any root system contains a subset (not uniquely defined) $\Sigma \subset \Phi$ of so-called *simple positive roots.* This is a set of d linearly independent roots $\Sigma = \{\alpha_1, \ldots, \alpha_d\}$ such that any root $\beta \in \Phi$ can be written either as a linear combination of α_j with non-negative integer coefficients, or as a linear combination with non-positive integer coefficients. We call Σ a *basis* of the root system Φ. A root system is conveniently represented by its *Dynkin diagram*. This is a graph with d nodes corresponding to the simple positive roots. Between two nodes j and k no line is drawn if the angle between α_j and α_k is $90°$, a single line if it is $120°$, a double line for $135°$ and a triple line for $150°$. It is only necessary to understand the geometry of *irreducible root systems*, where the Dynkin diagram is connected. Disconnected Dynkin diagrams (reducible root systems) are trivially understood in terms of products of irreducible root systems. For irreducible root systems, the roots are either all of the same length or have just two different lengths. In the latter case a marker $<$ or $>$ on an edge indicates the separation of long and short roots (short $<$ long). Since the work of W. Killing and E. Cartan in the late 19th century it has been known that Dynkin diagrams of irreducible root systems must belong to one of four possible infinite cases A_n $(n > 0)$, B_n $(n > 1)$, C_n $(n > 2)$, D_n $(n > 3)$ or five special cases E_6, E_7, E_8, F_4, G_2 shown in Figure 9.1. We say that two root systems are equivalent if they differ only by a scaling or an isometry. Up to equivalence there corresponds a unique root system to each Dynkin diagram.

A root system Φ is associated with a dual root system $\Phi^\vee$ defined such that a root $\alpha \in \Phi$ corresponds to a co-root $\alpha^\vee := 2\alpha/(\alpha,\alpha) \in \Phi^\vee$. If all roots have equal lengths then $\Phi^\vee = \Phi$ (up to equivalence), i.e. the root system is *self dual*. For the cases with two root lengths we have $B_2^\vee = B_2$, $B_n^\vee = C_n$ $(n > 2)$, $F_4^\vee = F_4$ and $G_2^\vee = G_2$.

Weyl groups and affine Weyl groups. Given a d-dimensional root system Φ with dual root system $\Phi^\vee$, for the roots $\alpha \in \Phi$, consider the reflection $s_\alpha \colon \mathcal{V} \to \mathcal{V}$ given by

$$s_\alpha(t) = t - \frac{2(t,\alpha)}{(\alpha,\alpha)}\alpha = t - (t,\alpha^\vee)\alpha.$$

For the dual roots $\alpha^\vee \in \Phi^\vee$, we define translations $\tau_{\alpha^\vee} : \mathcal{V} \to \mathcal{V}$ as

$$\tau_{\alpha^\vee}(t) = t + \alpha^\vee.$$

The *Weyl group* of Φ is the finite group of isometries on $\mathcal{V}$ generated by the reflections s_α for $\alpha \in \Sigma$:

$$W = \langle \{s_\alpha\}_{\alpha\in\Sigma} \rangle .$$

An important example of Weyl groups is the infinite family A_n. The Weyl group W of A_n is isomorphic to the symmetric group S_{n+1}. This is a group of order $|S_{n+1}| = (n+1)!$, consisting of all permutations of $n+1$ objects, and is also isomorphic to the symmetry group of the regular n-simplex. In particular the Weyl group of A_2 has order 6 and can be identified with the symmetries of the regular triangle.

The *dual root lattice* $L^\vee$ is the lattice spanned by the translations $\tau_{\alpha^\vee}$ for $\alpha^\vee \in \Sigma^\vee$. We identify this with the abelian group of translations on $\mathcal{V}$ generated by the dual roots

$$L^\vee = \langle \{\tau_{\alpha^\vee}\}_{\alpha^\vee \in \Sigma^\vee} \rangle .$$

The *affine Weyl group* $\widetilde{W}$ is the infinite crystallographic symmetry group of $\mathcal{V}$ generated by the reflections s_α for $\alpha \in \Sigma$ and the translations $\tau_{\alpha^\vee}$ for $\alpha^\vee \in \Sigma^\vee$, thus it is the semidirect product of the Weyl group W with the dual[5] root lattice $L^\vee$

$$\widetilde{W} = \langle \{s_\alpha\}_{\alpha \in \Sigma}, \{\tau_{\alpha^\vee}\}_{\alpha^\vee \in \Sigma^\vee} \rangle = W \rtimes L^\vee .$$

Let $\Lambda = (L^\vee)^\perp$ denote the reciprocal lattice of $L^\vee$. The lattice Λ is spanned by vectors $\{\lambda_j\}_{j=1}^d$ such that $(\lambda_j, \alpha_k^\vee) = \delta_{j,k}$ for all $\alpha_k^\vee \in \Sigma^\vee$. The vectors λ_j are called the *fundamental dominant weights* of the root system Φ, and Λ is called the *weights lattice*.

The *positive Weyl chamber* $\mathcal{C}_+$ is defined as the closed conic subset of $\mathcal{V}$ containing the points with nonnegative coordinates with respect to the dual basis $\{\lambda_1, \ldots, \lambda_d\}$, in other words

$$\mathcal{C}_+ = \{t \in \mathcal{V} \colon (t, \alpha_j) \geq 0\}.$$

This is a fundamental domain for the Weyl group acting on $\mathcal{V}$. The boundary of $\mathcal{C}_+$ consists of the hyperplanes perpendicular to $\{\alpha_1, \ldots, \alpha_d\}$. The affine Weyl group contains reflection symmetries about affine planes perpendicular to the roots, shifted a half integer multiple of the length of a co-root away from the origin, i.e. for each $\alpha^\vee \in \Phi^\vee$ and each $k \in Z$ there is an affine plane consisting of the points $P_{k,\alpha^\vee} = \{t \in \mathcal{V} \colon 2(t, \alpha^\vee) = k(\alpha^\vee, \alpha^\vee)\} = \{t \in \mathcal{V} \colon (t, \alpha) = k\}$, and this affine plane is invariant under the affine reflection $\tau_{k\alpha^\vee} \cdot s_\alpha$. A connected closed subset of $\mathcal{V}$ limited by such affine planes is called an *alcove* and is a fundamental domain for the affine Weyl group $\widetilde{W}$.

The situation is particularly simple for irreducible root systems, where the alcoves are always d-simplices. Recall that all roots $\alpha \in \Phi$ can be written as $\alpha = \sum_{k=1}^d n_k \alpha_k$ where all $n_k = 2(\alpha, \lambda_k)/(\alpha_k, \alpha_k)$ are either

[5] Since the Weyl groups of Φ and of $\Phi^\vee$ are identical, it is no problem to instead define $\widetilde{W} = W \rtimes L$ as the semidirect product of the Weyl group with the *primal* root lattice. We have, however, chosen to follow the most common definition here, which leads to a slightly simpler notation for the Fourier analysis.

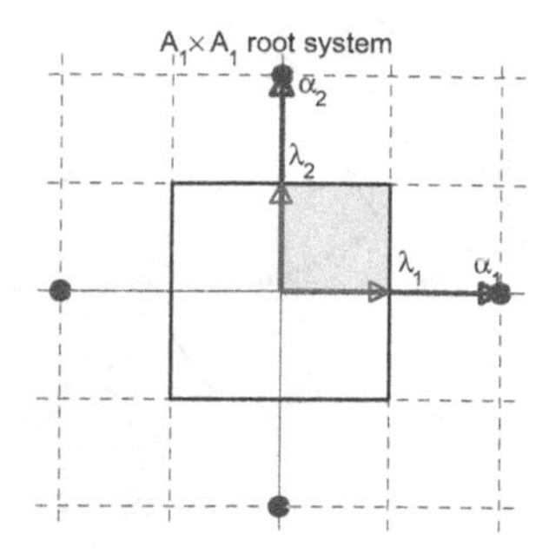

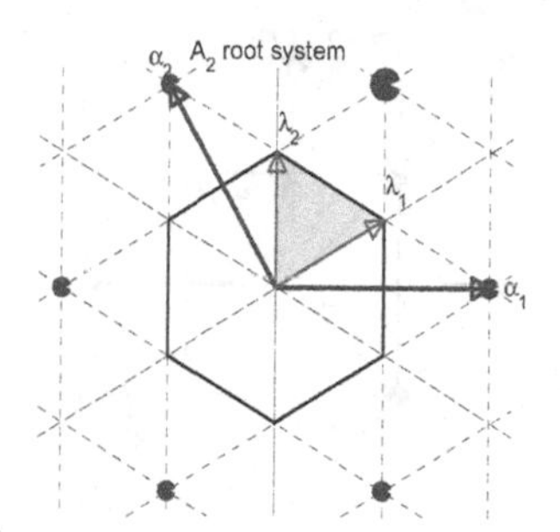

Figure 9.2 Reducible root system $A_1 \times A_1$ and irreducible system A_2

non-negative or all non-positive integers. A root $\tilde{\alpha}$ strictly dominates another root α, written $\tilde{\alpha} \succ \alpha$, if $\tilde{n}_k \geq n_k$ for all k, with strict inequality for at least one k. Irreducible root systems have a unique *dominant root* $\tilde{\alpha} \in \Phi$ such that $\tilde{\alpha} \succ \alpha$ for all $\alpha \neq \tilde{\alpha}$. The dominant root $\tilde{\alpha}$ is the unique long root in the Weyl chamber (possibly on the boundary). The basic geometric properties of affine Weyl groups are summarized by the following lemma:

Lemma 9.21

1 *If Φ is irreducible with dominant root $\tilde{\alpha}$ then a fundamental domain for $\widetilde{W}$ is the simplex $\triangle \subset \mathcal{V}$ given as*

$$\triangle = \{t \in \mathcal{V}\colon (t, \tilde{\alpha}) \leq 1 \text{ and } (t, \alpha_j) \geq 0 \text{ for all } \alpha_j \in \Sigma\},$$

where $\triangle$ has corners in the origin and in the points $\lambda_j/(\lambda_j, \tilde{\alpha})$ for $j = 1, \ldots, d$.

2 *The affine Weyl group is generated by the affine reflections about the boundary faces of the fundamental domain $\triangle$. For irreducible Φ these are*

$$\widetilde{W} = \langle \{s_{\alpha_j}\}_{j=1}^{d}, \tau_{\tilde{\alpha}} \cdot s_{\tilde{\alpha}} \rangle.$$

3 *If Φ is reducible then a fundamental domain for the affine Weyl group is given as the Cartesian product of the fundamental domains for each of its irreducible components.*

The simplest rank d root system is the reducible system $A_1 \times \cdots \times A_1$, where the Dynkin diagram consists of d non-connected dots. Figure 9.2 shows $A_1 \times A_1$. The solid black square is the fundamental domain of the root lattice, and the small shaded square the fundamental domain of the affine Weyl group $\widetilde{W}$.

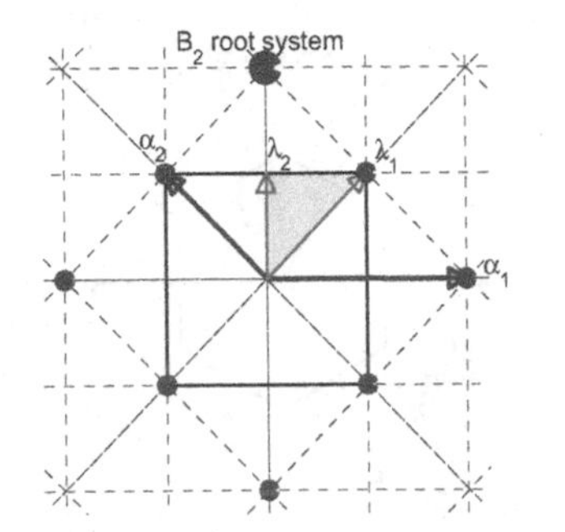

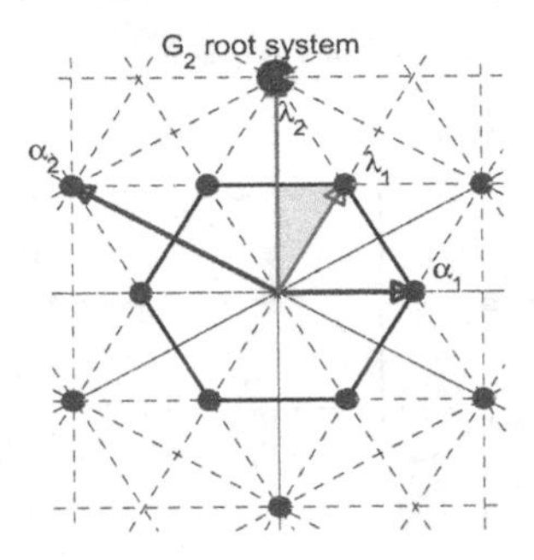

Figure 9.3 The irreducible root systems B_2 and G_2

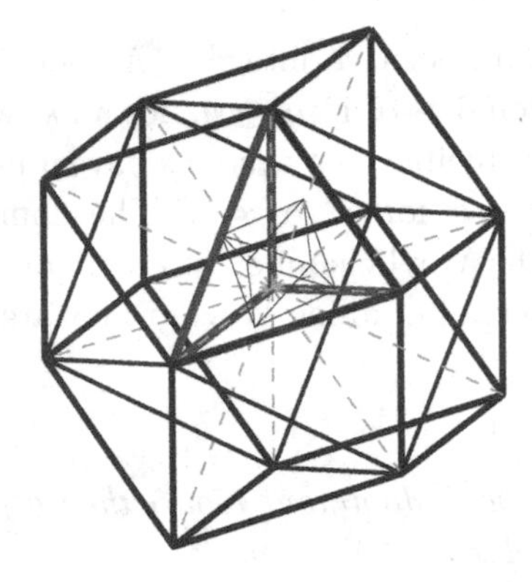

Figure 9.4 Fundamental domains of A_3 root system

The right part of Figure 9.2 and Figure 9.3 shows the irreducible 2-d cases with roots α (large dots), dominant root $\tilde{\alpha}$ (circle), simple positive roots (α_1, α_2), and fundamental dominant weights (λ_1, λ_2). The roots are normalized such that the longest roots have length $\sqrt{2}$, thus for long roots $\alpha^\vee = \alpha$. For short roots we have for B_2 that $\alpha^\vee = 2\alpha$ and for G_2 that $\alpha^\vee = 3\alpha$. The fundamental domain of the dual root lattice (Voronoi region of $L^\vee$) is indicated by ⬡, □ and ⬡, and the fundamental domain for the affine Weyl group (an alcove) is indicated by shaded triangles.

Figure 9.4 shows the A_3 case (self dual), where the fundamental domain (Voronoi region) of the root lattice is a rhombic dodecahedron, a convex polyhedron with 12 rhombic faces. Each of these faces (composed of two triangles) is part of a plane perpendicular to one of the 12 roots, halfway out to the root (roots are not drawn). The fundamental domain of the affine Weyl group is the tetrahedron with an inscribed octahedron. The corners of this tetrahedron constitute the origin and the fundamental dominant weights λ_1, λ_2, λ_3. The regular octahedron drawn inside the Weyl chamber is important for applications of multivariate Chebyshev polynomials in domain decompositions [9].

9.3.3 Laplacian eigenfunctions on triangles, tetrahedra and simplices

In this section we will consider real or complex valued functions on $\mathcal{V}$ respecting symmetries of an affine Weyl group. Consider first $L^2(\mathrm{T})$, the space of complex valued L^2-integrable periodic functions on the torus $\mathrm{T} = \mathcal{V}/L^\vee$, i.e. functions f such that $f(y + \alpha^\vee) = f(y)$ for all $y \in \mathcal{V}$ and $\alpha^\vee \in L^\vee$. Since the weights lattice Λ is reciprocal to $L^\vee$, the Fourier transform and its inverse are given for $f \in L^2(\mathrm{T})$ as

$$\widehat{f}(\lambda) = \mathcal{F}(f)(\lambda) = \frac{1}{\mathrm{vol}(\mathrm{T})}\int_{\mathrm{T}} f(t)\langle -\lambda, t\rangle dt, \tag{9.11}$$

$$f(t) = \mathcal{F}^{-1}(\widehat{f})(t) = \sum_{\lambda\in\Lambda} \widehat{f}(\lambda)\langle\lambda, t\rangle. \tag{9.12}$$

We are interested in functions which are both periodic under translations in $L^\vee$ and also respect the other symmetries in $\widetilde{W}$, e.g. functions with odd or even symmetry with respect to the reflections in $\widetilde{W}$[6]. Due to the semi-direct product structure $\widetilde{W} = W \rtimes L^\vee$ it follows that any $\tilde{w} \in \widetilde{W}$ can be written as $\tilde{w} = w \cdot \tau_{\alpha^\vee}$, where $w \in W$ and $\alpha^\vee \in L^\vee$. Thus on the space of periodic functions $L^2(\mathrm{T})$, the action of $\widetilde{W}$ and the finite group W are identical. We define subspaces of symmetric and skew-symmetric periodic functions as follows

$$L^2_\vee(\mathrm{T}) = \{f \in L^2(\mathrm{T})\colon f(wt) = f(t) \text{ for all } w \in W,\, t \in T\}$$
$$L^2_\wedge(\mathrm{T}) = \{f \in L^2(\mathrm{T})\colon f(wt) = (-1)^{|w|} f(t) \text{ for all } w \in W,\, t \in T\},$$

where $|w|$ denotes the *length*, defined as $|w| = \ell$, where $w = s_{\alpha_{j_1}} \cdots s_{\alpha_{j_\ell}}$ is written in the shortest possible way as a product of reflections about the simple positive roots $\alpha_j \in \Sigma$. Thus $(-1)^{|w|} = \det(w) = \pm 1$ depending on whether w is a product of an even or an odd number of reflections.

[6] There are other possible symmetries as well, related to other representations of the Weyl group [9]

We define L^2 orthogonal projections $\pi_\vee^W$ and $\pi_\wedge^W$ on these subspaces as

$$\pi_\vee^W f(t) = \frac{1}{|W|} \sum_{w \in W} f(wt), \tag{9.13}$$

$$\pi_\wedge^W f(t) = \frac{1}{|W|} \sum_{w \in W} (-1)^{|w|} f(wt). \tag{9.14}$$

Orthogonal bases for these subspaces are obtained by projecting the exponentials, yielding the cosine and sine type basis functions

$$c_\lambda(t) := \pi_\vee^W \exp 2\pi i(\lambda, t)$$
$$s_\lambda(t) := \pi_\wedge^W \exp 2\pi i(\lambda, t).$$

Note that these are not all distinct functions. Since they possess symmetries

$$c_\lambda(t) = c_{w\lambda}(t),$$
$$s_\lambda(t) = (-1)^{|w|} s_{w\lambda}(t),$$

for every $w \in W$, we need only one λ from each orbit of W. The weights in the Weyl chamber, $\Lambda_+ = \mathcal{C}_+ \cap \Lambda$, form a natural index set of orbit representatives, and we find L^2 orthogonal bases by taking the corresponding $c_\lambda(t)$ and $s_\lambda(t)$. Lemma 9.19 also holds in this more general case.

Lemma 9.22 *Given a rank d root system ϕ with weights lattice Λ, let W be the Weyl group and $\triangle$ denote the fundamental domain of the affine Weyl group $\widetilde{W} = W \rtimes L^\vee$. The functions $\{c_\lambda(t)\}_{\lambda \in \Lambda_+}$ and $\{s_\lambda(t)\}_{\lambda \in \Lambda_+}$ form two distinct L^2 orthogonal bases for $L^2(\triangle)$. These basis functions are eigenfunctions of the Laplacian ∇^2 on $\triangle$ satisfying homogeneous Neumann and Dirichlet boundary conditions, as in Lemma 9.19.*

Truncations of these bases do not, unfortunately, form spectrally convergent approximation spaces for analytic functions on $\triangle$. Approximation of a function f, defined on $\triangle$, in terms of $\{c_\lambda(t)\}$ is equivalent to Fourier approximation of the even extension of f in $L^2_\vee(\mathrm{T})$, and we do in general only observe quadratic convergence due to discontinuity of the gradient across the boundary of $\triangle$. A route to spectral convergence is by a change of variables which turns the trigonometric polynomials $\{c_\lambda(t)\}$ and $\{s_\lambda(t)\}$ into multivariate Chebyshev polynomials of first and second kind.

9.3.4 Multivariate Chebyshev polynomials

Recall that classical Chebyshev polynomials of first and second kind, $T_k(x)$ and $U_k(x)$, are obtained from $\cos(k\theta)$ and $\sin(k\theta)$ by a change of variable $x = \cos(\theta)$ as

$$T_k(x) = \cos(k\theta),$$
$$U_k(x) = \frac{\sin((k+1)\theta)}{\sin(\theta)}.$$

We want to understand this construction in the context of affine Weyl groups. We recognize $\cos(k\theta)$ and $\sin(k\theta)$ as the symmetrized and skew-symmetrized exponentials. The $\cos(\theta)$ used in the change of variables is the 2π-periodic function that is symmetric, non-constant and has the longest wavelength (as such, uniquely defined up to a constant). In other words $\cos(\theta) = \pi_\vee \exp(\lambda_1\theta)$, where $\lambda_1 = 1$ is the generator of the weights lattice. Any periodic band limited even function f has a symmetric Fourier series of finite support on the weights lattice, and must hence be a polynomial in the variable $x = \cos(\theta)$. The denominator $\sin(\theta)$ is similarly the odd function of longest possible wavelength. Any periodic band limited odd function f has a skew-symmetric Fourier series on the weights lattice. Dividing out by $\sin(\theta)$ results in a band limited even function which again must map to a polynomial under our change of variables. The denominator, which in this special case is $\sin(\theta)$, is called the *Weyl denominator*. It plays an important role in representation theory of compact Lie groups as the denominator in *Weyl's* character formula, a cornerstone of representation theory. We will detail these constructions in the sequel.

As before, we let Φ be a rank d root system on $\mathcal{V} = \mathbb{R}^d$, with Weyl group W, co-root lattice $L^\vee$ and affine Weyl group $\widetilde{W} = W \rtimes L^\vee$. Let $\mathrm{T} = \mathcal{V}/L^\vee$ be the torus of periodicity and $\Lambda = \mathrm{span}_{\mathbb{Z}}\{\lambda_1, \ldots, \lambda_d\}$ the reciprocal lattice of $L^\vee$. It is convenient to write the group Λ in multiplicative form, where we let $\{e^\lambda\}_{\lambda\in\Lambda}$ denote the elements of the multiplicative group, understood as formal symbols such that for $\lambda, \mu \in \Lambda$ we have $e^\lambda \cdot e^\mu = e^{\lambda+\mu}$.

Let $\mathcal{E} = \mathcal{E}(\mathbb{C}) \subset L^2(\Lambda)$ denote the free complex vector space over the symbols e^λ. This consists of all formal sums $a = \sum_{\lambda\in\Lambda} a(\lambda)e^\lambda$ where the coefficients $a(\lambda) \in \mathbb{C}$ and all but a finite number of these are non-zero. An element $a \in \mathcal{E}$ is identified with a trigonometric polynomial $f(t) = \mathcal{F}^{-1}(a)(t)$ on the torus T (i.e. a band limited periodic function) through the Fourier transforms given in (9.11)–(9.12).

Let $\mathcal{E}_\vee^W \subset \mathcal{E}$ denote the symmetric subalgebra of those elements that are invariant under the action of the Weyl group W on $\mathcal{E}$. This consists of those $a \in \mathcal{E}$ where $a(\lambda) = a(w\lambda)$ for all λ and all $w \in W$. Similarly, $\mathcal{E}_\wedge^W \subset \mathcal{E}$ denotes those $a \in \mathcal{E}$ that are alternating sign under reflections s_α, i.e. where the coefficients satisfy $a(\lambda) = (-1)^{|w|} a(w\lambda)$. Projections $\pi_\vee^W : \mathcal{E} \to \mathcal{E}_\vee$ and $\pi_\wedge^W : \mathcal{E} \to \mathcal{E}_\wedge$ are defined as in (9.13)–(9.14).

The algebra $\mathcal{E}$ is generated by $\{e^{\lambda_j}\}_{j=1}^d \cup \{e^{-\lambda_j}\}_{j=1}^d$, where λ_j are the fundamental dominant weights. $\mathcal{E}_\vee^W$ is the subalgebra generated by the symmetric generators $\{z_j\}_{j=1}^d$ defined as

$$z_j = \pi_\vee^W e^{\lambda_j} = \frac{1}{|W|} \sum_{w \in W} e^{w\lambda_j} = \frac{2}{|W|} \sum_{w \in W^+} e^{w\lambda_j}, \tag{9.15}$$

where W^+ denotes the even subgroup of W containing those w such that $|w|$ is even. The latter identity follows from $s_{\alpha_j}\lambda_j = \lambda_j$, thus it is enough to consider only w of even length. The action of W^+ on λ_j is free and effective.

It can be shown that $\mathcal{E}_\vee^W$ is a unique factorization domain over the generators $\{z_j\}$, i.e. any $a \in \mathcal{E}_\vee^W$ can be expressed uniquely as a polynomial in $\{z_j\}_{j=1}^d$.

The skew subspace $\mathcal{E}_\wedge^W$ does not form an algebra, but this can be corrected by dividing out the *Weyl denominator*. Define the *Weyl vector* $\rho \in \Lambda$ as

$$\rho = \sum_{j=1}^d \lambda_j = \frac{1}{2} \sum_{\alpha \in \Phi^+} \alpha.$$

We define the Weyl denominator $D \in \mathcal{E}_\wedge^W$ as

$$D = \sum_{w \in W} (-1)^{|w|} e^{w\rho}.$$

Proposition 9.23 *Any $a \in \mathcal{E}_\wedge^W$ is divisible by D, i.e. there exists a unique $b \in \mathcal{E}_\vee^W$ such that $a = bD$.*

Proof See [8] Prop. 25.2. □

Any $a \in \mathcal{E}_\vee^W$ can be written as a polynomial in $z_1, \ldots, z_d$, hence the following polynomials are well-defined.

Definition 9.24 For $\lambda \in \Lambda$ we define multivariate Chebyshev polynomials of first and second kind T_λ and U_λ as the unique polynomials that satisfy

$$T_\lambda(z_1,\ldots,z_d) = \pi_\vee^W e^\lambda = \frac{1}{|W|}\sum_{w\in W} e^{w\lambda}, \tag{9.16}$$

$$U_\lambda(z_1,\ldots,z_d) = \frac{|W|\pi_\wedge^W e^{\lambda+\rho}}{D} = \frac{\sum_{w\in W}(-1)^{|w|}e^{w(\rho+\lambda)}}{\sum_{w\in W}(-1)^{|w|}e^{w\rho}}. \tag{9.17}$$

By a slight abuse of notation, we will also consider z_j as W-invariant functions in $\mathbb{C}T$ as

$$z_j(t) = \mathcal{F}^{-1}(z_j)(t) = \frac{2}{|W|}\sum_{w\in W^+}\langle w\lambda_j, t\rangle_T.$$

The functions $z_j(t)$ may be real or complex. If there exists an $w \in W^+$ such that $w\lambda_j = \lambda_j$ then $z_j = \overline{z_j}$ is real. Otherwise there must exist an index $\overline{j} \neq j$ and a $w \in W^+$ such that $w\lambda_j = \lambda_{\overline{j}}$ and we have $\overline{z_j} = z_{\overline{j}}$. In the latter case we can replace these with d real coordinates $x_j = \frac{1}{2}(z_j + z_{\overline{j}})$, $x_{\overline{j}} = \frac{1}{2i}(z_j - z_{\overline{j}})$.

We remark that (9.17) is exactly the same formula as Weyl's character formula, giving the trace of all the irreducible characters on a semisimple Lie group [8]. These characters form an L^2 orthogonal basis for the space of class functions on the Lie group. Thus, expansions in terms of second kind multivariate Chebyshev polynomials are equivalent to expansions in terms of irreducible characters on a Lie group. In a similar way, the basis given by the irreducible representations block-diagonalize equivariant linear operators on a Lie group, it is also such that one may use the irreducible characters to obtain block diagonalizations, see [24]. Thus, our software, which is primarily constructed to deal with spectral element discretizations of PDEs, may also have important applications in computations on Lie groups. This opens up a whole area of possible applications of these approximations.

We will briefly summarize important properties of the multivariate Chebyshev polynomials which are presented in detail in [28, 29, 32, 9], and we will be a bit more detailed on some properties which are not detailed in these references.

Continuous orthogonality. Let Φ be an irreducible root system on $\mathcal{V} = \mathbb{R}^d$ with an alcove $\triangle$ being the simplex defined in Lemma 9.21. The corresponding family of multivariate Chebyshev polynomials are orthogonal on the domain

$$\delta = z(\triangle),$$

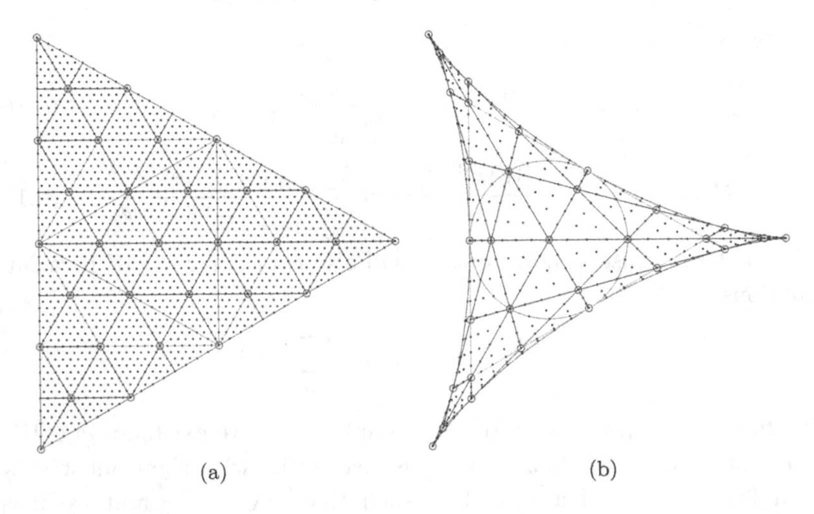

Figure 9.5 The equilateral domain Δ in (a) maps to the Deltoid δ in (b) under $t \mapsto z(t)$.

with respect to the inner product

$$(f, g) = \int_\delta \overline{f(z)} g(z) \frac{1}{\sqrt{D\overline{D}}} dz,$$

where D is the Weyl denominator. Figure 9.5 shows $\triangle$ and δ in the A_2 case. Here δ is a *deltoid*, a domain with cusps in each corner. For the A_3 case, $\triangle$ is shown in Figure 9.4 and δ in Figure 9.6. It poses a problem for applications that the domains δ are *not* simplices. However, in the A_2 case, there is a hexagon inside $\triangle$ which maps to an equilateral triangle inscribed in δ, whereas in the A_3 case there is an octagon in $\triangle$ which maps to a tetrahedron inscribed in δ. Note that the tetrahedron is not regular, but has two sides of length 1 and four sides of length $\sqrt{17/18}$. In our spectral element methods we have applied overlapping δ-subdomains such that the inscribed triangle and tetrahedron form a simplicial (non-overlapping) subdivision.

Discrete orthogonality and sampling. Pulled back to t-coordinates, polynomials in z become band limited symmetric functions $f \in L^2_\vee(\mathrm{T})$. Due to the Shannon sampling theorem, any band limited function $f \in L^2(\mathrm{T})$ can be exactly reconstructed from sampling on a sufficiently fine lattice $S < \mathrm{T}$. To preserve periodicity, we require that

$$S^\perp < \Lambda,$$

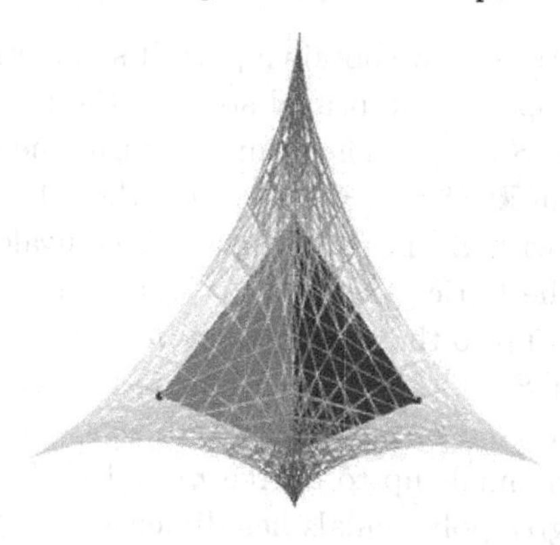

Figure 9.6 The deltoid domain δ in case A_3. The inscribed tetrahedron is the image of the regular octahedron inscribed in the Weyl chamber of $\triangle$, see Figure 9.4.

and in addition, we also require that the sampling lattice is W invariant,

$$WS = S.$$

Such a lattice S leads to exact discrete orthogonality formulae and discrete quadrature rules for polynomials such that $f(t)$ satisfies the Shannon criterion

$$\mathrm{supp}(\widehat{f}) \subset \widehat{\sigma}(\widehat{S}),$$

where $\widehat{\sigma}$ is a Voronoi transversal. There are several ways to construct such a lattice. In [29, 32] we work with down scalings of the co-root lattice,

$$S = \frac{1}{m} L^\vee,$$

where the integer m is chosen large enough for S to satisfy the Shannon criterion in the polynomial space where we want to perform approximations. This lattice is shown in Figure 9.5 in the A_2 case with circles for $m = 12$ and dots for $m = 48$. The resulting approximation space is the span of the multivariate polynomials up to degree $m/2$, except one particular polynomial of degree $m/2$ which aliases to 0 on S. In addition, the space contains some particular polynomials of degree up to $\frac{2}{3}m$.

Another lattice, which in many respects is more elegant, is the downscaling of the weights lattice

$$S = \frac{1}{m}\Lambda, \quad \text{for } m \in \mathbb{N}.$$

With this discretization we obtain a perfect symmetry between primal space and Fourier space. The primal space is the periodic domain $T = \mathbb{R}^n/L^\vee$, sampled in $S = \frac{1}{m}\Lambda$. The sampling turns the Fourier space into the periodic domain $\mathbb{R}^n/S^\perp = \mathbb{R}^n/mL^\vee$. On the other hand, periodicity under translation with $L^\vee$ in primal space is equivalent to sampling in Fourier space at the lattice $(L^\vee)^\perp = \Lambda$. Thus, the discretized periodic domain is self dual up to the scaling with m.

The sampling at $S = \frac{1}{m}\Lambda$ is perfect also in the sense that we, for the A_n Weyl groups, obtain an approximation space consisting of exactly the n-variate polynomials up to degree m and nothing else. The space of n-variate m-degree polynomials has dimension $\binom{m+n}{m}$. The alcove $\triangle$ of the affine Weyl group A_n is a simplex with corners in the origin and in λ_j, the principal dominant weights. Hence S is a downscaling of a simplex containing $\binom{m+n}{m}$ points, in particular these are the triangle numbers for $n = 2$ and the pyramidal numbers for $n = 3$. In Fourier space the situation is the same, the discrete space contains exactly the n-variate Chebyshev polynomials of degree up to and including m.

Lebesgue numbers. The elegant sampling results above are useless unless we can guarantee small Lebesque numbers and hence stable sampling. Fortunately, the following result holds for the A_n family sampled at the downscaled weights lattice.

Theorem 9.25 *Let λ be the weights lattice in the affine Weyl group A_n and let $S = \frac{1}{m}\Lambda$. Let $z\colon \triangle \to \delta$ be the coordinate change in* (9.15). *The Lebesgue number of the points $z(S)$ grows as*

$$\mathcal{O}(\log^n(m)).$$

Similar results hold for other lattices (e.g. downscaled roots lattice) and other affine Weyl groups. To prove this theorem, we must first show that the Lagrangian interpolating polynomials in the nodal set $z(S)$ are given as

$$\ell_j(z(t)) = c \sum_{w \in W} \mathcal{D}_S(w(t - t_j)) \quad \text{for all } t_j \in S,$$

where $\mathcal{D}_S(t)$ is the polyhedral Dirichlet kernel. The result follows from general bounds on Lebesgue numbers of the polyhedral Dirichlet kernels [34, 36]. We omit the details. Similar results are shown for special cases in [37].

Symmetric FFTs and fast Chebyshev expansions. In this paragraph we let Φ be the root system A_n with Weyl group $W \simeq S_{n+1}$ with $|W| = (n+1)!$. Let $S = \frac{1}{m}\Lambda$ and let G be the finite abelian group $G = S/L^\vee$. Under the action of W on G, the fundamental domain is $\triangle_m = \triangle \cup S$, where $\triangle$ is the n-simplex with corners in the origin and in the fundamental dominant weights λ_j. Thus $\triangle_m$ consists of $\binom{m+n}{m}$ points. The Weyl group also acts on $\widehat{G}$ via the adjoint $\langle \xi, wx \rangle = \langle \widehat{w}\xi, x \rangle$, with fundamental domain $\widehat{\triangle}_m$. Note that this case is self dual, so we can identify $\triangle_m$ and $\widehat{\triangle}_m$.

Let $p(z)$ be a real n-variate polynomial of degree m and let $f = p(z(\triangle_m))$ be the sampling of $p(z)$. The Fourier transform is defined as

$$\widehat{f}(\lambda) = \mathcal{F}_G(\pi^W_\vee f)(\lambda) \quad \text{for } \lambda \in \triangle_m.$$

The Chebyshev expansion of $p(z)$ is

$$p(z) = \sum_{\lambda \in \widehat{\triangle}_m} \widehat{f}(\lambda) T_\lambda(z).$$

How fast can this Chebyshev expansion be computed? One possibility is to extend f to $\mathbb{C}G$ and apply an FFT here. The number of points in G is approximately $N \approx (n+1)!\binom{m+n}{m}$. Counting both multiplications and additions, a complex FFT on N points costs $5N\log_2(N)$ floating point operations (flops).

It is, however, possible to improve the speed of this computation by a factor $2|W| = 2(n+1)!$ by using a symmetric FFT, as in [28]. The factor 2 comes from the assumption that $p(z)$ is real, not complex, and the factor $(n+1)!$ by carefully exploiting all the symmetries. This can be implemented as a Cooley–Tukey style FFT computation. Letting $m = 2^k$ we have a sequence of sub-lattices

$$S_1 < S_2 < S_4 < \cdots < S_{2^k} = S_m.$$

We know that Cooley–Tukey can be formulated with respect to arbitrary decompositions in sub-lattices. All the sub-lattices preserve all the symmetries of W, thus by organizing the computations carefully, it is sufficient to compute in the fundamental domains of the action of W. However, the Cooley–Tukey splitting is not just a splitting in sub-lattices, it also involves computations in the cosets of the sub-lattices. The cosets may lose some symmetries of W, and for such cosets the fundamental domain becomes larger. However, in that case the lost symmetry will map one coset to another coset, and we may discard the cosets that are equivalent to some other coset by a symmetry. In the dual space $\widehat{G}$, the situation is similar, W acts by the adjoint action and we can similarly exploit all the symmetries. In addition we have the real symmetry expressed as $\widehat{f}(\lambda) = \overline{\widehat{f}(-\lambda)}$.

To make a rather complicated story short, by carefully paying attention to exploiting all the symmetries, it can be shown:

Theorem 9.26 *Let π^n_m be the space of n-variate real polynomials of degree m, a real vector space of dimension $N = \binom{m+n}{m}$. The A_n Chebyshev expansion of $p(z) \in \pi^n_m$ can be computed in a stable manner by sampling $p(z)$ in the N points $z(\triangle_m)$. The cost of this computation is $\frac{5}{2}N\log_2(N)$ floating point numbers, counting both additions and multiplications.*

Recurrence formulae. Practical computations with Chebyshev expansions rely on the availability of fast transforms between 'nodal' and 'modal' representation, i.e. between sampled values and Chebyshev expansion coefficients. Products of polynomials can be computed fast in the nodal domain, while other operations such as integration and derivation can be done fast in the modal domain. Also, in the multivariate case, there exist recursion formulae similar to the univariate case.

From the convolution product in $\mathcal{E}^W_\vee$, one finds that the T_λ satisfy the recurrence relations

$$
\begin{aligned}
T_0 &= 1,\\
T_{\lambda_j} &= z_j,\\
T_\lambda &= T_{w\lambda} \text{ for } w \in W,\\
T_{-\lambda} &= \overline{T_\lambda},\\
T_\lambda T_\mu &= \frac{1}{|W|}\sum_{w\in W} T_{\lambda+w\mu}.
\end{aligned}
$$

These reduce to classical three-term recurrences for A_1 and four term recurrences for A_2, see [29].

There also exist linear recurrence relations which compute the Chebyshev expansion of $\nabla p(z)$ from the Chebyshev expansion of $p(z)$, see [32].

9.4 Conclusions

We have discussed various aspects of group theory as a tool for analyzing numerical algorithms and for developing new computational algorithms. In particular, we have focused on algorithms related to sampling in regular lattices and algorithms related to applications of reflection groups and kaleidoscopes. Group theory provides an important conceptual framework both for understanding the fundamental structure of the algorithms and also for structuring the implementation of numerical software.

Connections between group theory and Chebyshev approximations has been discussed in detail. This is a connection which goes much deeper than what has been presented in this paper. Weyl's character formula for the irreducible characters on simple Lie groups is essentially identical to the definition of second kind multivariate Chebyshev polynomials. Chebyshev approximation theory is therefore intimately linked to approximation theory on Lie groups.

Another topic, which has not been discussed in this paper, is the application of group theory in the development of algorithms for computing time evolution of dynamical systems. Lie group integrators are general numerical methods built around Lie group actions on the phase space of the dynamical system [23]. Lie group integrators rely upon exact computations of operator exponentials. The interplay between Lie group integrators for time evolution and Chebyshev based spectral element methods for spatial discretizations is an active area of research where all these techniques come together.

References

[1] K. Åhlander, M. Haveraaen and H. Munthe-Kaas, On the role of mathematical abstractions for scientific computing. In *The Architecture of Scientific Software*, R. F. Boisvert and P. T. P. Tang, Editors, IFIP Advances in Information and Communication Technology, **60**, 145–158, 2001.

[2] K. Åhlander and H. Munthe-Kaas, Applications of the generalized Fourier transform in numerical linear algebra. *BIT Numerical Mathematics*, **45**, (2005), 819–850.

[3] E. L. Allgower, K. Georg, R. Miranda and J. Tausch, Numerical exploitation of equivariance. *ZAMM*, **78**, (1998), 185–201.

[4] L. Auslander, *Lecture Notes on Nil-Theta Functions.* Regional Conference Series in Mathematics, **34**, AMS, 1977.

[5] L. Auslander and R. Tolimieri, Is computing with the finite Fourier transform pure or applied mathematics. *Notices AMS*, **1**, 1979.

[6] R. J. Beerends, Chebyshev polynomials in several variables and the radial part of the Laplace–Beltrami operator. *Trans. AMS*, **328**, (1991), 779–814.

[7] A. Bossavit, Symmetry, groups, and boundary value problems. a progressive introduction to noncommutative harmonic analysis of partial differen-

tial equations in domains with geometrical symmetry. *Comput. Methods Appl. Mech. Engrg.*, **56**, (1986), 167–215.

[8] D. Bump, *Lie Groups.* Springer Verlag, 2004.

[9] S. H. Christiansen, H. Z. Munthe-Kaas and B. Owren, Topics in structure-preserving discretization. *Acta Numerica*, **20**, (2011), 1–119.

[10] J. H. Conway, N. J. A. Sloane and E. Bannai, *Sphere Packings, Lattices, and Groups.* Springer Verlag, 1999.

[11] C. C. Douglas and J. Mandel, Abstract theory for the domain reduction method. *Computing*, **48**, (1992), 73–96.

[12] M. Dubiner, Spectral methods on triangles and other domains. *J. Scientific Computing*, **6**, (1991), 345–390.

[13] R. Eier and R. Lidl, A class of orthogonal polynomials in k variables. *Math. Ann.*, **260**, (1982), 93–99.

[14] K. Engo, A. Marthinsen and H. Z. Munthe-Kaas, Diffman: An object-oriented matlab toolbox for solving differential equations on manifolds. *Applied Numer. Math.*, **39**, (2001), 323–347.

[15] A. F. Fässler and E. Stiefel, *Group Theoretical Methods and their Applications.* Birkhäuser, Boston, 1992.

[16] K. Georg and R. Miranda, Exploiting symmetry in solving linear equations. In *Bifurcation and Symmetry*, E. L. Allgower, K. Böhmer and M. Golubisky, editors, **104** of *ISNM*, 157–168, Birkhäuser, Basel, 1992.

[17] F. X. Giraldo and T. Warburton, A nodal triangle-based spectral element method for the shallow water equations on the sphere. *J. Comput. Phys.*, **207**, (2005), 129-150.

[18] R. Hadani and A. Singer, Representation theoretic patterns in three dimensional cryo-electron microscopy I: the intrinsic reconstitution algorithm. *Annals Math.*, **174**, (2011), 1219–1241.

[19] R. Hadani and A. Singer, Representation theoretic patterns in three-dimensional cryo-electron microscopy II—the class averaging problem. *Found. Comput. Math.*, **11**, (2011), 589–616.

[20] J. S. Hesthaven and T. Warburton, *Nodal Discontinuous Galerkin Methods: Algorithms, Analysis, and Applications.* Springer-Verlag, New York, 2008.

[21] M. E. Hoffman and W. D. Withers, Generalized Chebyshev polynomials associated with affine Weyl groups. *Trans. AMS*, **308**, (1988), 91–104.

[22] D. Huybrechs, On the Fourier extension of non-periodic functions. *SIAM J. Numer. Anal.*, **47**, (2010), 4326–4355.

[23] A. Iserles, H. Munthe-Kaas, S. P. Nørsett and A. Zanna, Lie-group methods. *Acta Numerica*, **9**, (2000), 215–365.

[24] G. James and M. Liebeck, *Representations and Characters of Groups.* Cambridge University Press, 2nd edition, 2001.

[25] T. Koornwinder, Orthogonal polynomials in two variables which are eigenfunctions of two algebraically independent partial differential operators I–IV. *Indiag. Math.*, **36**, (1974), 48–66 and 357–381.

[26] R. Lidl, Tchebyscheffpolynome in mehreren Variabelen. *J. Reine Angew. Math*, **273**, (1975), 178–198.

[27] J. S. Lomont, *Applications of Finite Groups*. Academic Press, New York, 1959.

[28] H. Munthe-Kaas, Symmetric FFTs; a general approach. In *Topics in linear algebra for vector- and parallel computers, PhD thesis*. NTNU, Trondheim, Norway, 1989. Available at: `http://hans.munthe-kaas.no`.

[29] H. Z. Munthe-Kaas, On group Fourier analysis and symmetry preserving discretizations of PDEs. *J. Physics A: Mathematical and General*, **39**, (2006), 5563.

[30] H. Munthe-Kaas and T. Sørevik, Multidimensional pseudo-spectral methods on lattice grids. To appear in *Applied Numerical Mathematics*.

[31] W. Rudin, *Fourier Analysis on Groups*. Wiley-Interscience, 1990.

[32] B. N. Ryland and H. Z. Munthe-Kaas, On multivariate Chebyshev polynomials and spectral approximations on triangles. In *Spectral and High Order Methods for Partial Differential Equations*, J. S. Hesthaven and E. M. Rønquist, Editors, Lecture Notes Comp. Sci. and Eng., **76**, 19–41, 2011.

[33] J. P. Serre, *Linear Representations of Finite Groups*. Springer, 1977.

[34] R. J. Stanton and P. A. Tomas, Polyhedral summability of Fourier series on compact Lie groups. *Amer. J. Math.*, **100**, (1978), 477–493.

[35] S. Thangavelu, *Harmonic Analysis on the Heisenberg Group*. Birkhauser, 1998.

[36] G. Travaglini, Polyhedral summability of multiple Fourier series. *Colloq. Math*, **65**, (1993), 103–116.

[37] Y. Xu, Fourier series and approximation on hexagonal and triangular domains. *Constr. Approx.*, **31**, 2010, 115–138.

10

Sage: Creating a Viable Free Open Source Alternative to Magma, Maple, Mathematica, and MATLAB

William Stein[a]
Department of Mathematics
University of Washington

Abstract

Sage is a large free open source software package aimed at all areas of mathematical computation. Hundreds of people have contributed to the project, which has steadily grown in popularity since 2005. This paper describes the motivation for starting Sage and the history of the project.

10.1 Introduction

The goal of the Sage project (`http://www.sagemath.org`) is to create a viable free open source alternative to Magma, Maple™, Mathematica®, and MATLAB®, which are the most popular non-free closed source mathematical software systems.[1] Magma is (by far) the most advanced non-free system for structured abstract algebraic computation, Mathematica and Maple are popular and highly developed systems that shine at symbolic manipulation, and MATLAB is the most popular system for applied numerical mathematics. Together there are over 3,000 employees working at the companies that produce the four Ma's listed above, which take in over a hundred million dollars of revenue annually.

By a viable free alternative to the Ma's, we mean a system that will have the important mathematical features of each Ma, with comparable speed. It will have 2d and 3d graphics, an interactive graphical user interface, and documentation, including books, papers, school and col-

[a] The work was supported by NSF grant DUE-1022574.

[1] Maple is a trademark of Waterloo Maple Inc. Mathematica is a registered trademark of Wolfram Research Incorporated. MATLAB is a registered trademark of MathWorks. I will refer to the four systems together as "the Ma's" in the rest of this article.

lege curriculum materials, etc. A single alternative to all of the Ma's is not necessarily a drop-in replacement for any of the Ma's; in particular, it need not run programs written in the custom languages of those systems. Thus an alternative may be philosophically different than the open source system Octave, which understands the MATLAB source language and attempts to implement the entire MATLAB library. Development could instead focus on implementing functions that users demand, rather than systematically trying to implement every single function of the Ma's. The culture, architecture, and general look and feel of such a system would be very different than that of the Ma's.

In Section 10.2 we explain some of the motivation for starting the Sage project, in Section 10.3 we describe the basic architecture of Sage, and in Section 10.4 we sketch aspects of the history of the project.

10.2 Motivation for starting Sage

Each of the Ma's cost substantial money, and is hence expensive for me, my collaborators, and students. The Ma's are not *owned by the community* like Sage is, or Wikipedia is, for that matter.

The Ma's are closed, which means that the implementation of some algorithms are secret, in which case you are not allowed to modify or extend them.

> "You should realize at the outset that while knowing about the internals of Mathematica may be of intellectual interest, it is usually much less important in practice than you might at first suppose. Indeed, in almost all practical uses of Mathematica, issues about how Mathematica works inside turn out to be largely irrelevant. Particularly in more advanced applications of Mathematica, it may sometimes seem worthwhile to try to analyze internal algorithms in order to predict which way of doing a given computation will be the most efficient. [...] But most often the analyses will not be worthwhile. For the internals of Mathematica are quite complicated.."
> – The Mathematica Documentation

The philosophy espoused in Sage, and indeed by the vast open source software community, is exactly the opposite. We want you to know about the internals, and when they are quite complicated, we want you to help make them more understandable. Indeed, Sage's growth depends on *you* analyzing how Sage works, improving it, and contributing your improvements back.

```
sage: crt(2, 1, 3, 5)   # Chinese Remainder Theorem
11
sage: crt?             # ? = documentation and examples
Returns a solution to a Chinese Remainder Theorem...
...
sage: crt??            # ?? = source code
def crt(...):
...
    g, alpha, beta = XGCD(m, n)
    q, r = (b - a).quo_rem(g)
    if r != 0:
        raise ValueError("No solution ...")
    return (a + q*alpha*m) % lcm(m, n)
```

Moreover, by browsing `http://hg.sagemath.org/sage-main/`, you can see exactly who wrote or modified any particular line of code in the Sage library, when they did it, and why. Everything included in Sage is free and open source, and it will forever remain that way.

> "I see open source as Science. If you don't spread your ideas in the open, if you don't allow other people to look at how your ideas work and verify that they work, you are not doing Science, you are doing Witchcraft. Traditional software development models, where you keep things inside a company and hide what you are doing, are basically Witchcraft. Open source is all about the fact that it is open; people can actually look at what you are doing, and they can improve it, and they can build on top of it. [...] One of my favorite quotes from history is Newton: 'If I had seen further, it has been by standing on the shoulders of giants.'"
> – Linus Torvalds.
> Listen at `http://www.youtube.com/watch?v=bt_Y4pSdsHw`

The design decisions of the Ma's are not made openly by the community. In contrast, important decisions about Sage development are made via open public discussions and voting that is archived on public mailing lists with thousands of subscribers.

Every one of the Ma's uses a special mathematics-oriented interpreted programming language, which locks you into their product, makes writing some code outside mathematics unnecessarily difficult, and impacts the number of software engineers that are experts at programming in that language. In contrast, the user language of Sage is primarily the mainstream free open source language Python `http://python.org`, which is one of the world's most popular interpreted programming languages. The Sage project neither invented nor maintains the underlying Python language, but gains immediate access to the IPython shell, Python scientific libraries (such as NumPy, SciPy, CVXopt and

MatPlotLib), and a large Python community with major support from big companies such as Google. In comparison to Python, the Ma's are small players in terms of language development. Thus for Sage most of the problems of language development are handled by someone else.

The bug tracking done for three of four of the Ma's is currently secret[2], which means that there is no published accounting of all known bugs, the status of work on them, and how bugs are resolved. But the Ma's do have many bugs; see the release notes of each new version, which lists bugs that were fixed[3]. Sage also has bugs, which are all publicly tracked at `http://trac.sagemath.org`, and there are numerous "Bug Days" workshops devoted entirely to fixing bugs in Sage. Moreover, all discussion about resolving a given bug, including peer review of solutions, is publicly archived. We note that sadly even some prize winning[4] free open source systems, such as GAP `http://www.gap-system.org/`, do not have an open bug tracking system, resulting in people reporting the same bugs over and over again.

Each of the Ma's is a combination of secret unchangeable compiled code and less secret interpreted code. Users with experience programming in compiled languages such as Fortran or C++ may find the loss of a compiler to be frustrating. None of the Ma's has an optimizing compiler that converts programs written in their custom interpreted language to a fast executable binary format that is not interpreted at runtime.[5] In contrast, Sage is tightly integrated with Cython[6] `http://www.cython.org`, which is a ython-to-C/C++ compiler that speeds up code execution and has support for statically declaring data types (for potentially enormous speedups) and natively calling existing C/C++/Fortran code. For example, enter the following in a cell of the Sage notebook (e.g., `http://sagenb.org`):

```
def python_sum2(n):
    s = int(0)
    for i in xrange(1, n+1):
        s += i*i
    return s
```

[2] MATLAB has an open bug tracker, though it requires free registration to view.
[3] See also `http://cybertester.com/` and `http://maple.bug-list.org/`.
[4] Jenks Prize, 2008
[5] MATLAB has a compiler, but "the source code is still interpreted at run-time, and performance of code should be the same whether run in standalone mode or in MATLAB." Mathematica also has a `Compile` function, but simply compiles expressions to a different internal format that is interpreted, much like Sage's `fast_callable` function.
[6] The Cython project has received extensive contributions from Sage developers, and is very popular in the world of Python-based scientific computing.

Then enter the following in another cell:

```
%cython
def cython_sum2(long n):
    cdef long i, s = 0
    for i in range(1, n+1):
        s += i*i
    return s
```

The second implementation, despite looking nearly identical, is nearly a hundred times faster than the first one (your timings may vary).

```
sage: timeit('python_sum2(2*10^6)')
5 loops, best of 3: 154 ms per loop
sage: timeit('cython_sum2(2*10^6)')
125 loops, best of 3: 1.76 ms per loop
sage: 154/1.76
87.5
```

Of course, it is better to choose a different algorithm. In case you don't remember a closed form expression for the sum of the first n squares, Sage can deduce it:

```
sage: var('k, n')
sage: factor(sum(k^2, k, 1, n))
1/6*(n + 1)*(2*n + 1)*n
```

And now our simpler fast implementation is:

```
def sum2(n):
    return n*(2*n+1)*(n+1)/6
```

Just as above, we can also use the Cython compiler:

```
%cython
def c_sum2(long n):
    return n*(2*n+1)*(n+1)/6
```

Comparing times, we see that Cython is 10 times faster:

```
sage: n = 2*10^6
sage: timeit('sum2(n)')
625 loops, best of 3: 1.41 microseconds per loop
sage: timeit('c_sum2(n)')
625 loops, best of 3: 0.145 microseconds per loop
sage: 1.41/.145
9.72413793103448
```

In this case, the enhanced speed comes at a cost, in that the answer is *wrong* when the input is large enough to cause an overflow:

```
sage: c_sum2(2*10^6)    # WARNING: overflow
-4077886789512586O3
```

Cython is very powerful, but to fully benefit from it, one must understand machine level arithmetic data types, such as long, int, float, etc. With Sage you have that option.

10.3 What is Sage?

The goal of Sage is to compete with the Ma's, and the intellectual property at our disposal is the complete range of GPL-compatibly licensed open source software.

Sage is a self-contained free open source *distribution* of about 100 open source software packages and libraries[7] that aims to address all computational areas of pure and applied mathematics. The download of Sage contains all dependencies required for the normal functioning of Sage, including Python itself. Sage includes a substantial amount of code that provides a unified Python-based *interface* to these other packages. Sage also includes a library of new code written in Python, Cython and C/C++, which implements a huge range of algorithms.

10.4 History

I made the first release of Sage in February 2005, and at the time called it "**S**oftware for **A**rithmetic **G**eometry **E**xperimentation." I was a serious user of, and contributor to, Magma at the time, and was motivated to start Sage for many of the reasons discussed above. In particular, I was personally frustrated with the top-down closed development model of Magma, the fact that *several million lines* of the source code of Magma are closed source, and the fees that my colleagues had to pay in order to use the substantial amount of code that I contributed to Magma. Despite my early naive hope that Magma would be open sourced, it never was. So I started Sage motivated by the dream that someday the single most

[7] See the list of packages in Sage at http://sagemath.org/packages/standard/. The list includes R, Pari, Singular, GAP, Maxima, GSL, Numpy, Scipy, ATLAS, Matplotlib, and many other popular programs.

important item of software I use on a daily basis would be free and open. David Joyner, David Kohel, Joe Wetherell, and Martin Albrecht were also involved in the development of Sage during the first year.

In February 2006, the National Science Foundation funded a 2-day workshop called "Sage Days 2006" at UC San Diego, which had about 40 participants and speakers from several open and closed source mathematical software projects. After doing a year of fulltime mostly solitary work on Sage, I was surprised by the positive reception of Sage by members of the mathematical research community. What Sage promised was something many mathematicians wanted. Whether or not Sage would someday deliver on that promise was (and for many still is) an open question.

I had decided when I started Sage that I would make it powerful enough for my research, with or without the help of anybody else, and was pleasantly surprised at this workshop to find that many other people were interested in helping, and understood the shortcomings of existing open source software, such as GAP and PARI, and the longterm need to move beyond Magma. Six months later, I ran another Sage Days workshop, which resulted in numerous talented young graduate students, including David Harvey, David Roe, Robert Bradshaw, and Robert Miller, getting involved in Sage development. I used startup money from University of Washington to hire Alex Clemesha as a fulltime employee to implement 2d graphics and help create a notebook interface to Sage. I also learned that there was much broader interest in such a system, and stopped referring to Sage as being exclusively for "arithmetic geometry"; instead, Sage became "**S**oftware for **A**lgebra and **G**eometry **E**xperimentation." Today the acronym is deprecated.

The year 2007 was a major turning point for Sage. Far more people got involved with development, we had four Sage Days workshops, and prompted by Craig Citro, we instituted a requirement that all new code must have tests for 100% of the functions touched by that code, and every modification to Sage must be peer reviewed. Our peer review process is much more open than in mathematical research journals; everything that happens is publicly archived at `http://trac.sagemath.org`. During 2007, I also secured some funding for Sage development from Microsoft Research, Google, and NSF. Also, a German graduate student studying cryptography, Martin Albrecht presented Sage at the Trophées du Libre competition in France, and Sage won first place in "Scientific Software", which led to a huge amount of good publicity, including articles in many languages around the world and appearances[8] on the front page of `http://slashdot.org`.

In 2008, I organized 7 Sage Days workshops at places such as IPAM (at UCLA) and the Clay Mathematics Institute, and for the first time, several people besides me made releases of Sage. In 2009, we had 8 more Sage Days workshops, and the underlying foundations of Sage improved, including development of a powerful coercion architecture. This *coercion model* systematically determines what happens when performing operations such as `a + b`, when `a` and `b` are elements of potentially different rings (or groups, or modules, etc.).

```
sage: R.<x> = PolynomialRing(ZZ)
sage: f = x + 1/2; f
x + 1/2
sage: parent(f)
Univariate Polynomial Ring in x over Rational Field
```

We compare this with Magma (V2.17-4), which has a more ad hoc coercion system:

```
> R<x> := PolynomialRing(IntegerRing());
> x + 1/2
    ^
Runtime error in '+': Bad argument types
Argument types given: RngUPolElt[RngInt], FldRatElt
```

Robert Bradshaw and I also added support for beautiful browser-based 3D graphics to Sage, which involved writing a 3D graphics library, and adapting the free open source JMOL Java library (see `http://jmol.sourceforge.net/`) for rendering molecules to instead plot mathematical objects.

```
sage: f(x,y) = sin(x - y) * y * cos(x)
sage: plot3d(f, (x,-3,3), (y,-3,3), color='red')
```

[8] For example, `http://science.slashdot.org/story/07/12/08/1350258/Open-Source-Sage-Takes-Aim-at-High-End-Math-Software`

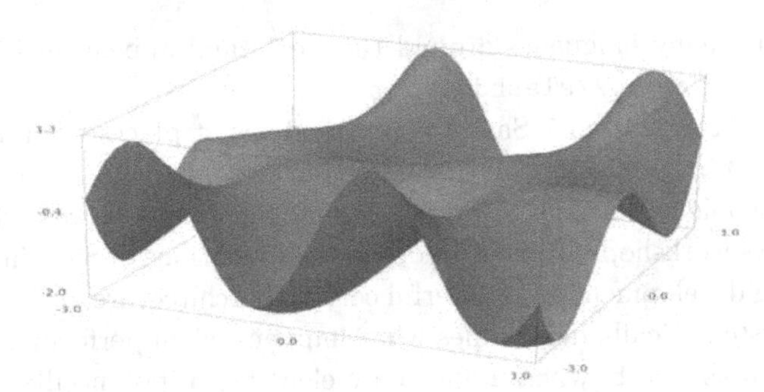

In 2009, following a huge amount of porting work by Mike Hansen, development of algebraic combinatorics in Sage picked up substantial momentum, with the switch of the entire MuPAD-combinat group to Sage (forming sage-combinat `http://wiki.sagemath.org/combinat`), only months before the formerly free system MuPAD®[9] was bought out by Mathworks (makers of MATLAB). In addition to work on Lie theory by Dan Bump, this also led to a massive amount of work on a category theoretic framework for Sage by Nicolas Thiery.

In 2010, there were 13 Sage Days workshops in many parts of the world, and grant funding for Sage significantly improved, including new NSF funding for undergraduate curriculum development. I also spent much of my programming time during 2010–2011 developing a number theory library called psage `http://code.google.com/p/purplesage/`, which is currently not included in Sage, but can be easily installed.

Many aspects of Sage make it an ideal tool for teaching mathematics, so there's a steadily growing group of teachers using it: for example, there have been MAA PREP workshops on Sage for the last two years, and a third is likely to run next summer, there are regular posts on the Sage lists about setting up classroom servers, and there is an NSF-funded project called UTMOST (see `http://utmost.aimath.org/`) devoted to creating undergraduate curriculum materials for Sage.

The page `http://sagemath.org/library-publications.html` lists 101 accepted publications that use Sage, 47 preprints, 22 theses, and 16 books, and there are surely many more "in the wild" that we are not aware of. According to Google Analytics, the main Sage website gets about 2,500 absolute unique visitors per day, and the website `http://sagenb.org`, which allows anybody to easily use Sage through their web browser, has around 700 absolute unique visitors per day.

For many mathematicians and students, Sage is today the mature, open source, and free foundation on which they can build their research program.

[9] MuPAD is a registered trademark of SciFace Software GmbH & Co.

Printed in the United States
by Baker & Taylor Publisher Services

Printed in the United States
by Baker & Taylor Publisher Services